NONRESIDENT
TRAINING
COURSE

SEPTEMBER 1998

Navy Electricity and Electronics Training Series

Module 8—Introduction to Amplifiers

NAVEDTRA 14180

DISTRIBUTION STATEMENT A: Approved for public release; distribution is unlimited.

Although the words "he," "him," and "his" are used sparingly in this course to enhance communication, they are not intended to be gender driven or to affront or discriminate against anyone.

DISTRIBUTION STATEMENT A: Approved for public release; distribution is unlimited.

PREFACE

By enrolling in this self-study course, you have demonstrated a desire to improve yourself and the Navy. Remember, however, this self-study course is only one part of the total Navy training program. Practical experience, schools, selected reading, and your desire to succeed are also necessary to successfully round out a fully meaningful training program.

COURSE OVERVIEW: To introduce the student to the subject of Amplifiers who needs such a background in accomplishing daily work and/or in preparing for further study.

THE COURSE: This self-study course is organized into subject matter areas, each containing learning objectives to help you determine what you should learn along with text and illustrations to help you understand the information. The subject matter reflects day-to-day requirements and experiences of personnel in the rating or skill area. It also reflects guidance provided by Enlisted Community Managers (ECMs) and other senior personnel, technical references, instructions, etc., and either the occupational or naval standards, which are listed in the Manual of Navy Enlisted Manpower Personnel Classifications and Occupational Standards, NAVPERS 18068.

THE QUESTIONS: The questions that appear in this course are designed to help you understand the material in the text.

VALUE: In completing this course, you will improve your military and professional knowledge. Importantly, it can also help you study for the Navy-wide advancement in rate examination. If you are studying and discover a reference in the text to another publication for further information, look it up.

1998 Edition Prepared by AY CM Keith E. Glading

Published by NAVAL EDUCATION AND TRAINING PROFESSIONAL DEVELOPMENT AND TECHNOLOGY CENTER

NAVSUP Logistics Tracking Number 0504-LP-026-8330

Sailor's Creed

"I am a United States Sailor.

I will support and defend the Constitution of the United States of America and I will obey the orders of those appointed over me.

I represent the fighting spirit of the Navy and those who have gone before me to defend freedom and democracy around the world.

I proudly serve my country's Navy combat team with honor, courage and commitment.

I am committed to excellence and the fair treatment of all."

TABLE OF CONTENTS

CHAPTER PAGE

1. Amplifiers 1-1
2. Video and Rf Amplifiers 2-1
3. Special Amplifiers 3-1

APPENDIX
I. Glossary AI-1
INDEX INDEX-1

NAVY ELECTRICITY AND ELECTRONICS TRAINING SERIES

The Navy Electricity and Electronics Training Series (NEETS) was developed for use by personnel in many electrical- and electronic-related Navy ratings. Written by, and with the advice of, senior technicians in these ratings, this series provides beginners with fundamental electrical and electronic concepts through self-study. The presentation of this series is not oriented to any specific rating structure, but is divided into modules containing related information organized into traditional paths of instruction.

The series is designed to give small amounts of information that can be easily digested before advancing further into the more complex material. For a student just becoming acquainted with electricity or electronics, it is highly recommended that the modules be studied in their suggested sequence. While there is a listing of NEETS by module title, the following brief descriptions give a quick overview of how the individual modules flow together.

Module 1, Introduction to Matter, Energy, and Direct Current, introduces the course with a short history of electricity and electronics and proceeds into the characteristics of matter, energy, and direct current (dc). It also describes some of the general safety precautions and first-aid procedures that should be common knowledge for a person working in the field of electricity. Related safety hints are located throughout the rest of the series, as well.

Module 2, Introduction to Alternating Current and Transformers, is an introduction to alternating current (ac) and transformers, including basic ac theory and fundamentals of electromagnetism, inductance, capacitance, impedance, and transformers.

Module 3, Introduction to Circuit Protection, Control, and Measurement, encompasses circuit breakers, fuses, and current limiters used in circuit protection, as well as the theory and use of meters as electrical measuring devices.

Module 4, Introduction to Electrical Conductors, Wiring Techniques, and Schematic Reading, presents conductor usage, insulation used as wire covering, splicing, termination of wiring, soldering, and reading electrical wiring diagrams.

Module 5, Introduction to Generators and Motors, is an introduction to generators and motors, and covers the uses of ac and dc generators and motors in the conversion of electrical and mechanical energies.

Module 6, Introduction to Electronic Emission, Tubes, and Power Supplies, ties the first five modules together in an introduction to vacuum tubes and vacuum-tube power supplies.

Module 7, Introduction to Solid-State Devices and Power Supplies, is similar to module 6, but it is in reference to solid-state devices.

Module 8, Introduction to Amplifiers, covers amplifiers.

Module 9, Introduction to Wave-Generation and Wave-Shaping Circuits, discusses wave generation and wave-shaping circuits.

Module 10, Introduction to Wave Propagation, Transmission Lines, and Antennas, presents the characteristics of wave propagation, transmission lines, and antennas.

Module 11, Microwave Principles, explains microwave oscillators, amplifiers, and waveguides. Module 12, Modulation Principles, discusses the principles of modulation.

Module 13, Introduction to Number Systems and Logic Circuits, presents the fundamental concepts of number systems, Boolean algebra, and logic circuits, all of which pertain to digital computers.

Module 14, Introduction to Microelectronics, covers microelectronics technology and miniature and microminiature circuit repair.

Module 15, Principles of Synchros, Servos, and Gyros, provides the basic principles, operations, functions, and applications of synchro, servo, and gyro mechanisms.

Module 16, Introduction to Test Equipment, is an introduction to some of the more commonly used test equipments and their applications.

Module 17, Radio-Frequency Communications Principles, presents the fundamentals of a radio-frequency communications system.

Module 18, Radar Principles, covers the fundamentals of a radar system.

Module 19, The Technician's Handbook, is a handy reference of commonly used general information, such as electrical and electronic formulas, color coding, and naval supply system data.

Module 20, Master Glossary, is the glossary of terms for the series.

Module 21, Test Methods and Practices, describes basic test methods and practices.

Module 22, Introduction to Digital Computers, is an introduction to digital computers.

Module 23, Magnetic Recording, is an introduction to the use and maintenance of magnetic recorders and the concepts of recording on magnetic tape and disks.

Module 24, Introduction to Fiber Optics, is an introduction to fiber optics.

Embedded questions are inserted throughout each module, except for modules 19 and 20, which are reference books. If you have any difficulty in answering any of the questions, restudy the applicable section.

Although an attempt has been made to use simple language, various technical words and phrases have necessarily been included. Specific terms are defined in Module 20, Master Glossary.

Considerable emphasis has been placed on illustrations to provide a maximum amount of information. In some instances, a knowledge of basic algebra may be required.

Assignments are provided for each module, with the exceptions of Module 19, The Technician's Handbook; and Module 20, Master Glossary. Course descriptions and ordering information are in NAVEDTRA 12061, Catalog of Nonresident Training Courses.

Throughout the text of this course and while using technical manuals associated with the equipment you will be working on, you will find the below notations at the end of some paragraphs. The notations are used to emphasize that safety hazards exist and care must be taken or observed.

WARNING

AN OPERATING PROCEDURE, PRACTICE, OR CONDITION, ETC.,

WHICH MAY RESULT IN INJURY OR DEATH IF NOT CAREFULLY OBSERVED OR FOLLOWED.

CAUTION

AN OPERATING PROCEDURE, PRACTICE, OR CONDITION, ETC., WHICH MAY RESULT IN DAMAGE TO EQUIPMENT IF NOT CAREFULLY OBSERVED OR FOLLOWED.

NOTE

An operating procedure, practice, or condition, etc., which is essential to emphasize.

INSTRUCTIONS FOR TAKING THE COURSE

ASSIGNMENTS

The text pages that you are to study are listed at the beginning of each assignment. Study these pages carefully before attempting to answer the questions. Pay close attention to tables and illustrations and read the learning objectives. The learning objectives state what you should be able to do after studying the material. Answering the questions correctly helps you accomplish the objectives.

SELECTING YOUR ANSWERS

Read each question carefully, then select the BEST answer. You may refer freely to the text. The answers must be the result of your own work and decisions. You are prohibited from referring to or copying the answers of others and from giving answers to anyone else taking the course.

SUBMITTING YOUR ASSIGNMENTS

To have your assignments graded, you must be enrolled in the course with the Nonresident Training Course Administration Branch at the Naval Education and Training Professional Development and Technology Center (NETPDTC). Following enrollment, there are two ways of having your assignments graded: (1) use the Internet to submit your assignments as you complete them, or (2) send all the assignments at one time by mail to NETPDTC.

Grading on the Internet: Advantages to Internet grading are:

• you may submit your answers as soon as you complete an assignment, and

• you get your results faster; usually by the next working day (approximately 24 hours).

In addition to receiving grade results for each assignment, you will receive course completion confirmation once you have completed all the

assignments. To submit your assignment answers via the Internet, go to:

http ://courses.cnet.na vy.mil

Grading by Mail: When you submit answer sheets by mail, send all of your assignments at one time. Do NOT submit individual answer sheets for grading. Mail all of your assignments in an envelope, which you either provide yourself or obtain from your nearest Educational Services Officer (ESO). Submit answer sheets to:

COMMANDING OFFICER NETPDTC N331 6490 SAUFLEY FIELD ROAD PENSACOLA FL 32559-5000

Answer Sheets: All courses include one "scannable" answer sheet for each assignment. These answer sheets are preprinted with your SSN, name, assignment number, and course number. Explanations for completing the answer sheets are on the answer sheet.

Do not use answer sheet reproductions: Use only the original answer sheets that we provide—reproductions will not work with our scanning equipment and cannot be processed.

Follow the instructions for marking your answers on the answer sheet. Be sure that blocks 1, 2, and 3 are filled in correctly. This information is necessary for your course to be properly processed and for you to receive credit for your work.

COMPLETION TIME

Courses must be completed within 12 months from the date of enrollment. This includes time required to resubmit failed assignments.

PASS/FAIL ASSIGNMENT PROCEDURES

If your overall course score is 3.2 or higher, you will pass the course and will not be required to resubmit assignments. Once your assignments have been graded you will receive course completion confirmation.

If you receive less than a 3.2 on any assignment and your overall course score is below 3.2, you will be given the opportunity to resubmit failed assignments. You may resubmit failed assignments only once. Internet students will receive notification when they have failed an assignment—they may then resubmit failed assignments on the web site. Internet students may view and print results for failed assignments from the web site. Students who submit by mail will receive a failing result letter and a new answer sheet for resubmission of each failed assignment.

COMPLETION CONFIRMATION

After successfully completing this course, you will receive a letter of completion.

ERRATA

Errata are used to correct minor errors or delete obsolete information in a course. Errata may also be used to provide instructions to the student. If a course has an errata, it will be included as the first page(s) after the front cover. Errata for all courses can be accessed and viewed/downloaded at:

http://www.advancement.cnet.navy.mil

STUDENT FEEDBACK QUESTIONS

We value your suggestions, questions, and criticisms on our courses. If you would like to communicate with us regarding this course, we encourage you, if possible, to use e-mail. If you write or fax, please use a copy of the Student Comment form that follows this page.

For subject matter questions:

E-mail: n315.products@cnet.navy.mil Phone: Comm: (850) 452-1001, ext. 1728 DSN: 922-1001, ext. 1728 FAX: (850)452-1370 (Do not fax answer sheets.) Address: COMMANDING OFFICER NETPDTC N315 6490 SAUFLEY FIELD ROAD PENSACOLA FL 32509-5237

For enrollment, shipping, grading, or completion letter questions

E-mail: fleetservices @ cnet. navy. mil

Phone: Toll Free: 877-264-8583

Comm: (850)452-1511/1181/1859 DSN: 922-1511/1181/1859 FAX: (850)452-1370 (Do not fax answer sheets.)

Address: COMMANDING OFFICER NETPDTC N331 6490 SAUFLEY FIELD ROAD PENSACOLA FL 32559-5000

NAVAL RESERVE RETIREMENT CREDIT

If you are a member of the Naval Reserve, you will receive retirement points if you are authorized to receive them under current directives governing retirement of Naval Reserve personnel. For Naval Reserve retirement, this course is evaluated at 4 points. (Refer to Administrative Procedures for Naval Reservists on Inactive Duty, BUPERSINST 1001.39, for more information about retirement points.)

Student Comments

NEETS Module 8 Course Title: Introduction to Amplifiers

NAVEDTRA: 14180 Date:

We need some information about you :

Rate/Rank and Name: SSN: Command/Unit

Street Address: City: State/FPO: Zip

Your comments, suggestions, etc.:

Privacy Act Statement: Under authority of Title 5, USC 301, information regarding your military status is requested in processing your comments and in preparing a reply. This information will not be divulged without written authorization to anyone other than those within POD for official use in determining performance.

NETPDTC 1550/41 (Rev 4-00)

CHAPTER 1

AMPLIFIERS

LEARNING OBJECTIVES

Learning objectives are stated at the beginning of each chapter. These learning objectives serve as a preview of the information you are expected to learn in the chapter. The comprehensive check questions are based on the objectives. By successfully completing the OCC/ECC, you indicate that you have met the objectives and have learned the information. The learning objectives are listed below.

Upon completion of this chapter, you will be able to:

1. Define amplification and list several common uses; state two ways in which amplifiers are classified.

2. List the four classes of operation of, four methods of coupling for, and the impedance characteristics of the three configurations of a transistor amplifier.

3. Define feedback and list the two types of feedback.

4. Describe and state one use for a phase splitter.

5. State a common use for and one advantage of a push-pull amplifier.

INTRODUCTION

This chapter is a milestone in your study of electronics. Previous modules have been concerned more with individual components of circuits than with the complete circuits as the subject. This chapter and the other chapters of this module are concerned with the circuitry of amplifiers. While components are discussed, the discussion of the components is not an explanation of the working of the component itself (these have been covered in previous modules) but an explanation of the component as it relates to the circuit.

The circuits this chapter is concerned with are AMPLIFIERS. Amplifiers are devices that provide AMPLIFICATION. That doesn't explain much, but it does describe an amplifier if you know what amplification is and what it is used for.

WHAT IS AMPLIFICATION?

Just as an amplifier is a device that provides amplification, amplification is the process of providing an increase in AMPLITUDE. Amplitude is a term that describes the size of a signal. In terms of a.c, amplitude usually refers to the amount of voltage or current. A 5-volt peak-to-peak a.c.signal would be larger in amplitude than a 4-volt peak-to-peak a.c. signal. "SIGNAL" is a general term used to refer to any a.c. or d.c. of interest in a circuit; e.g., input signal and output signal. A signal can be large or small, ac. or d.c, a sine wave or nonsinusoidal, or even nonelectrical such as sound or light. "Signal" is a very general term and, therefore, not very descriptive by itself, but it does sound more technical than the word "thing". It is not very impressive to refer to the "input thing" or the "thing that comes out of this circuit."

Perhaps the concept of the relationship of amplifier-amplification-amplitude will be clearer if you consider a parallel situation (an analogy). A magnifying glass is a magnifier. As such, it provides magnification which is an increase in the magnitude (size) of an object. This relationship of magnifier-magnification-magnitude is the same as the relationship of amplifier-amplification-amplitude. The analogy is true in one other aspect as well. The magnifier does not change the object that is being magnified; it is only the image that is larger, not the object itself. With the amplifier, the output signal differs in amplitude from the input signal, but the input signal still exists unchanged. So, the object (input signal) and the magnifier (amplifier) control the image (output signal).

An amplifier can be defined as a device that enables an input signal to control an output signal. The output signal will have some (or all) of the characteristics of the input signal but will generally be larger than the input signal in terms of voltage, current, or power.

USES OF AMPLIFICATION

Most electronic devices use amplifiers to provide various amounts of signal amplification. Since most signals are originally too small to control or drive the desired device, some amplification is needed.

For example, the audio signal taken from a record is too small to drive a speaker, so amplification is needed. The signal will be amplified several times between the needle of the record player and the speaker. Each time the signal is amplified it is said to go through a STAGE of amplification. The audio amplifier shown connected between the turntable and speaker system in figure 1-1 contains several stages of amplification.

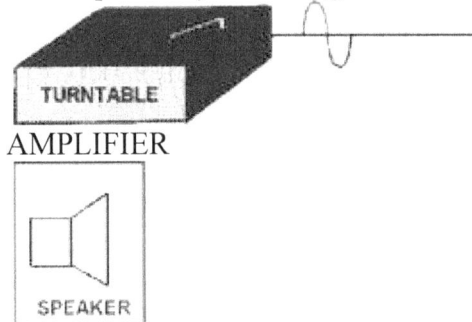

Figure 1-1.—Amplifier as used with turntable and speaker.

Notice the triangle used in figure 1-1 to represent the amplifier. This triangle is the standard block diagram symbol for an amplifier.

Another example of the use of an amplifier is shown in figure 1-2. In a radio receiver, the signal picked up by the antenna is too weak (small) to be used as it is. This signal must be amplified before it is sent to the detector. (The detector separates the audio signal from the frequency that was sent by the transmitter. The way in which this is done will be discussed later in this training series.)

ANTENNA

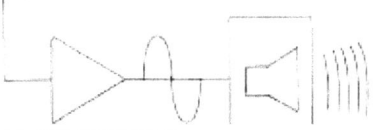

DETECTOR (SEPARATES AUDIO FROM RADIO FREQUENCY)
AMPLIFIER (RADIO FREQUENCY)

AMPLIFIER I
(AUDIO FREQUENCY) SPEAKER

Figure 1-2.—Amplifiers as used in radio receiver.

The audio signal from the detector will then be amplified to make it large enough to drive the speaker of the radio.

Almost every electronic device contains at least one stage of amplification, so you will be seeing amplifiers in many devices that you work on. Amplifiers will also be used in most of the NEETS modules that follow this one.

Q-1. What is amplification?

Q-2. Does an amplifier actually change an input signal? Why or why not? Q-3. Why do electronic devices use amplifiers?

CLASSIFICATION OF AMPLIFIERS

Most electronic devices use at least one amplifier, but there are many types of amplifiers. This module will not try to describe all the different types of amplifiers. You will be shown the general principles of amplifiers and some typical amplifier circuits.

Most amplifiers can be classified in two ways. The first classification is by their function. This means they are basically voltage amplifiers or power amplifiers. The second classification is by their frequency response. In other words what frequencies are they designed to amplify?

If you describe an amplifier by these two classifications (function and frequency response) you will have a good working description of the amplifier. You may not know what the exact circuitry is, but you will know what the amplifier does and the frequencies that it is designed to handle.

VOLTAGE AMPLIFIERS AND POWER AMPLIFIERS

All amplifiers are current-control devices. The input signal to an amplifier controls the current output of the amplifier. The connections of the amplifying device (electron tube, transistor, magnetic amplifier,

etc.) and the circuitry of the amplifier determine the classification. Amplifiers are classified as voltage or power amplifiers.

A VOLTAGE AMPLIFIER is an amplifier in which the output signal voltage is larger than the input signal voltage. In other words, a voltage amplifier amplifies the voltage of the input signal.

A POWER AMPLIFIER is an amplifier in which the output signal power is greater than the input signal power. In other words, a power amplifier amplifies the power of the input signal. Most power amplifiers are used as the final amplifier (stage of amplification) and control (or drive) the output device. The output device could be a speaker, an indicating device, an antenna, or the heads on a tape recorder. Whatever the device, the power to make it work (or drive it) comes from the final stage of amplification which is a power amplifier.

Figure 1-3 shows a simple block diagram of a voltage amplifier with its input and output signals and a power amplifier with its input and output signals. Notice that in view (A) the output signal voltage is larger than the input signal voltage. Since the current values for the input and output signals are not shown, you cannot tell if there is a power gain in addition to the voltage gain.

IN
INPUTSIGNAL 200 mV PEAK-TO-PEAK

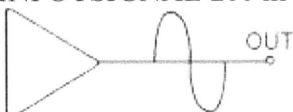

OUTPUT SIGNAL 4 V PEAK-TO-PEAK

Figure 1-3 A.—Block diagram of voltage and power amplifiers.

INPUTSIGNAL OUTPUT SIGNAL
10 V PEAK-TO-PEAK 5 V PEAK-TO-PEAK
100 mA(1W OF POWER) 2 A (10 W OF POWER)

Figure 1-3B.—Block diagram of voltage and power amplifiers.

In view (B) of the figure the output signal voltage is less than the input signal voltage. As a voltage amplifier, this circuit has a gain of less than 1. The output power, however, is greater than the input power. Therefore, this circuit is a power amplifier.

The classification of an amplifier as a voltage or power amplifier is made by comparing the characteristics of the input and output signals. If the output signal is larger in voltage amplitude than the input signal, the amplifier is a voltage amplifier. If there is no voltage gain, but the output power is greater than the input power, the amplifier is a power amplifier.

FREQUENCY RESPONSE OF AMPLIFIERS

In addition to being classified by function, amplifiers are classified by frequency response. The frequency response of an amplifier refers to the band of frequencies or frequency range that the amplifier was designed to amplify.

You may wonder why the frequency response is important. Why doesn't an amplifier designed to amplify a signal of 1000 Hz work just as well at 1000 MHz? The answer is that the components of the amplifier respond differently at different frequencies. The amplifying device (electron tube, transistor, magnetic amplifier, etc.) itself will have frequency limitations and respond in different ways as the frequency changes. Capacitors and inductors in the circuit will change their reactance as the frequency changes. Even the slight amounts of capacitance and inductance between the circuit wiring and other components (interelectrode capacitance and self-inductance) can become significant at high frequencies. Since the response of components varies with the frequency, the components of an amplifier are selected to amplify a certain range or band of frequencies.

NOTE: For explanations of interelectrode capacitance and self-inductance see

NEETS Modules 2 — Introduction to Alternating Current and Transformers; 6—Introduction to Electronic Emission, Tubes, and Power Supplies; and 7— Introduction to Solid-State Devices and Power Supplies.

The three broad categories of frequency response for amplifiers are AUDIO AMPLIFIER, RF AMPLIFIER, and VIDEO AMPLIFIER.

An audio amplifier is designed to amplify frequencies between 15 Hz and 20 kHz. Any amplifier that is designed for this entire band of frequencies or any band of frequencies contained in the audio range is considered to be an audio amplifier.

In the term rf amplifier, the "rf stands for radio frequency. These amplifiers are designed to amplify frequencies between 10 kHz and 100,000 MHz. A single amplifier will not amplify the entire rf range, but any amplifier whose frequency band is included in the rf range is considered an rf amplifier.

A video amplifier is an amplifier designed to amplify a band of frequencies from 10 Hz to 6 MHz. Because this is such a wide band of frequencies, these amplifiers are sometimes called WIDE-BAND AMPLIFIERS. While a video amplifier will amplify a very wide band of frequencies, it does not have the gain of narrower-band amplifiers. It also requires a great many more components than a narrow-band amplifier to enable it to amplify a wide range of frequencies.

Q-4. In what two ways are amplifiers classified?

Q-5. What type of amplifier would be used to drive the speaker system of a record player? Q-6. What type of amplifier would be used to amplify the signal from a radio antenna?

TRANSISTOR AMPLIFIERS

A transistor amplifier is a current-control device. The current in the base of the transistor (which is dependent on the emitter-base bias) controls the current in the collector. A vacuum-tube amplifier is also a current-control device. The grid bias controls the plate current. These facts are expanded upon in NEETS Module 6, Introduction to Electronic Emission, Tubes and Power Supplies, and Module 7, Introduction to Solid-State Devices and Power Supplies.

You might hear that a vacuum tube is a voltage-operated device (since the grid does not need to draw current) while the transistor is a current-operated device. You might agree with this statement, but both the vacuum tube and the transistor are still current-control devices. The whole secret to understanding amplifiers is to remember that fact. Current control is the name of the game. Once current is controlled you can use it to give you a voltage gain or a power gain.

This chapter will use transistor amplifiers to present the concepts and principles of amplifiers. These concepts apply to vacuum-tube amplifiers and, in most cases, magnetic amplifiers as well as transistor amplifiers. If you wish to study the vacuum-tube equivalent circuits of the transistor circuits presented, an excellent source is the EIMB, NAVSEA 0967-LP-000-0120, Electronics Circuits.

The first amplifier concept that is discussed is the "class of operation" of an amplifier.

AMPLIFIER CLASSES OF OPERATION

The class of operation of an amplifier is determined by the amount of time (in relation to the input signal) that current flows in the output circuit. This is a function of the operating point of the amplifying device. The operating point of the amplifying device is determined by the bias applied to the device. There are four classes of operation for an amplifier. These are: A, AB, B and C. Each class of operation has certain uses and characteristics. No one class of operation is "better" than any other class. The selection of the "best" class of operation is determined by the use of the amplifying circuit. The best class of operation for a phonograph is not the best class for a radio transmitter.

Class A Operation

A simple transistor amplifier that is operated class A is shown in figure 1-4. Since the output signal is a 100% (or 360°) copy of the input signal, current in the output circuit must flow for 100% of the input signal time. This is the definition of a class A amplifier. Amplifier current flows for 100% of the input signal.

The class A amplifier has the characteristics of good FIDELITY and low EFFICIENCY. Fidelity means that the output signal is just like the input signal in all respects except amplitude. It has the same

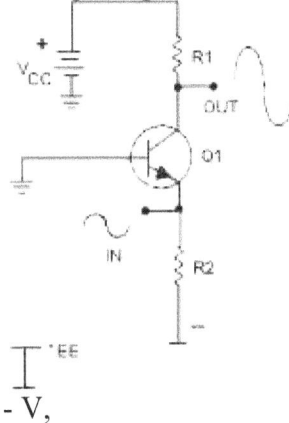

- V,

Figure 1-4.—A simple class A transistor amplifier.

shape and frequency. In some cases, there may be a phase difference between the input and output signal (usually 180°), but the signals are still considered to be "good copies." If the output signal is not like the input signal in shape or frequency, the signal is said to be DISTORTED. DISTORTION is any undesired change in a signal from input to output.

The efficiency of an amplifier refers to the amount of power delivered to the output compared to the power supplied to the circuit. Since every device takes power to operate, if the amplifier operates for 360° of input signal, it uses more power than if it only operates for 180° of input signal. If the amplifier uses more power, less power is available for the output signal and efficiency is lower. Since class A amplifiers operate (have current flow) for 360° of input signal, they are low in efficiency. This low efficiency is acceptable in class A amplifiers because they are used where efficiency is not as important as fidelity.

Class AB Operation

If the amplifying device is biased in such a way that current flows in the device for 51% - 99% of the input signal, the amplifier is operating class AB. A simple class AB amplifier is shown in figure 1-5.

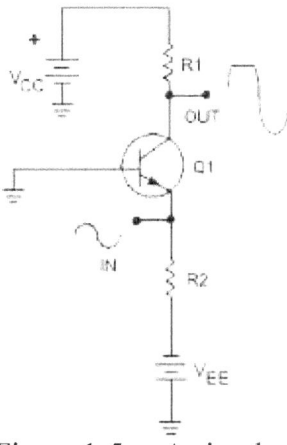

Figure 1-5.—A simple class AB transistor amplifier.

Notice that the output signal is distorted. The output signal no longer has the same shape as the input signal. The portion of the output signal that appears to be cut off is caused by the lack of current through the transistor. When the emitter becomes positive enough, the transistor cannot conduct because the base-to-emitter junction is no longer forward biased. Any further increase in input signal will not cause an increase in output signal voltage.

Class AB amplifiers have better efficiency and poorer fidelity than class A amplifiers. They are used when the output signal need not be a complete reproduction of the input signal, but both positive and negative portions of the input signal must be available at the output.

Class AB amplifiers are usually defined as amplifiers operating between class A and class B because class A amplifiers operate on 100% of input signal and class B amplifiers (discussed next) operate on 50% of the input signal. Any amplifier operating between these two limits is operating class AB.

Class B Operation

As was stated above, a class B amplifier operates for 50% of the input signal. A simple class B amplifier is shown in figure 1-6.

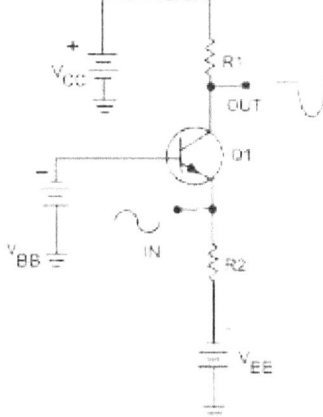

Figure 1-6.—A simple class B transistor amplifier.

In the circuit shown in figure 1-6, the base-emitter bias will not allow the transistor to conduct whenever the input signal becomes positive. Therefore, only the negative portion of the input signal is reproduced in the output signal. You may wonder why a class B amplifier would be used instead of a simple rectifier if only half the input signal is desired in the output. The answer to this is that the rectifier does not amplify. The output signal of a rectifier cannot be higher in amplitude than the input signal. The class B amplifier not only reproduces half the input signal, but amplifies it as well.

Class B amplifiers are twice as efficient as class A amplifiers since the amplifying device only conducts (and uses power) for half of the input signal. A class B amplifier is used in cases where exactly 50% of the input signal must be amplified. If less than 50% of the input signal is needed, a class C amplifier is used.

Class C Operation

Figure 1-7 shows a simple class C amplifier. Notice that only a small portion of the input signal is present in the output signal. Since the transistor does not conduct except during a small portion of the input signal, this is the most efficient amplifier. It also has the worst fidelity. The output signal bears very little resemblance to the input signal.

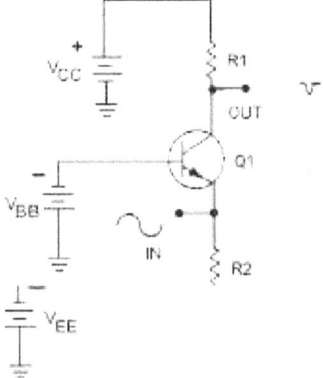

Figure 1-7.—A simple class C transistor amplifier.

Class C amplifiers are used where the output signal need only be present during part of one-half of the input signal. Any amplifier that operates on less than 50% of the input signal is operated class C.

Q-7. What determines the class of operation of an amplifier?

Q-8. What are the four classes of operation of a transistor amplifier?

Q-9. If the output of a circuit needs to be a complete representation of one-half of the input signal, what class of operation is indicated?

Q-10. Why is class C operation more efficient than class A operation?

Q-11. What class of operation has the highest fidelity?

AMPLIFIER COUPLING

Earlier in this module it was stated that almost every electronic device contains at least one stage of amplification. Many devices contain several stages of amplification and therefore several amplifiers. Stages of amplification are added when a single stage will not provide the required amount of amplification. For example, if a single stage of amplification will provide a maximum gain of 100 and the desired gain from the device is 1000, two stages of amplification will be required. The two stages might have gains of 10 and 100, 20 and 50, or 25 and 40. (The overall gain is the product of the individual stages-10 x 100 = 20 x 50 = 25 x 40 = 1000.)

Figure 1-8 shows the effect of adding stages of amplification. As stages of amplification are added, the signal increases and the final output (from the speaker) is increased.

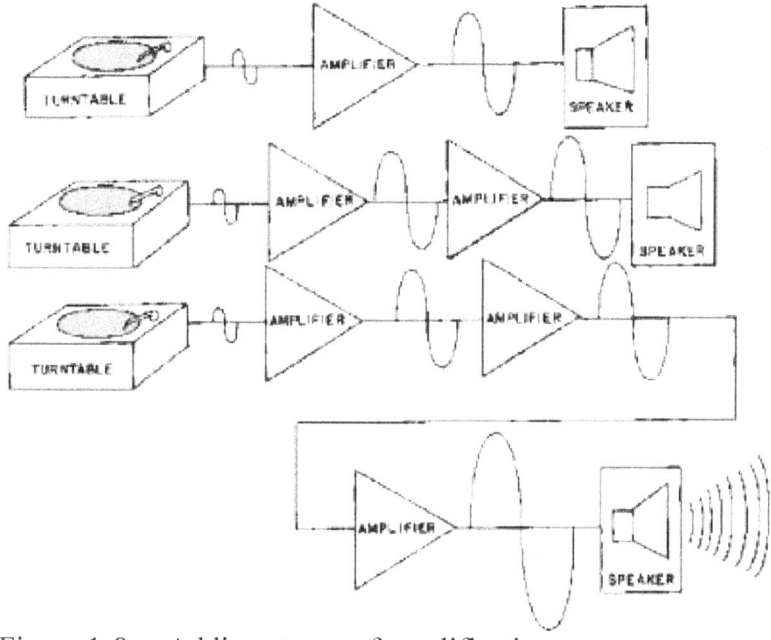

Figure 1-8.—Adding stages of amplification.

Whether an amplifier is one of a series in a device or a single stage connected between two other devices (top view, figure 1-8), there must be some way for the signal to enter and leave the amplifier. The process of transferring energy between circuits is known as COUPLING. There are various ways of coupling signals into and out of amplifier circuits. The following is a description of some of the more common methods of amplifier coupling.

Direct Coupling

The method of coupling that uses the least number of circuit elements and that is, perhaps, the easiest to understand is direct coupling. In direct coupling the output of one stage is connected directly to the input of the following stage. Figure 1-9 shows two direct-coupled transistor amplifiers.

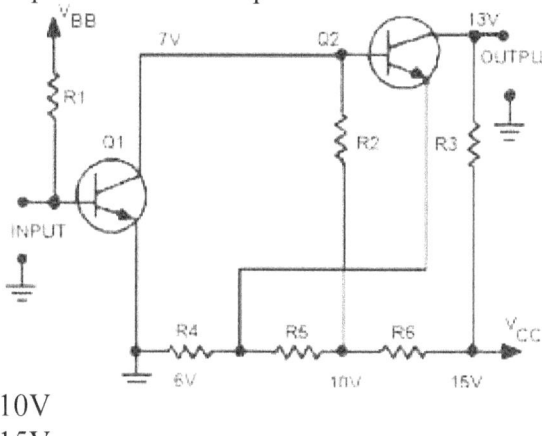

10V
15V

Figure 1-9.—Direct-coupled transistor amplifiers.

Notice that the output (collector) of Q1 is connected directly to the input (base) of Q2. The network of R4, R5, and R6 is a voltage divider used to provide the bias and operating voltages for Q1 and Q2. The entire circuit provides two stages of amplification.

Direct coupling provides a good frequency response since no frequency-sensitive components (inductors and capacitors) are used. The frequency response of a circuit using direct coupling is affected only by the amplifying device itself.

Direct coupling has several disadvantages, however. The major problem is the power supply requirements for direct-coupled amplifiers. Each succeeding stage requires a higher voltage. The load and voltage divider resistors use a large amount of power and the biasing can become very complicated. In addition, it is difficult to match the impedance from stage to stage with direct coupling. (Impedance matching is covered a little later in this chapter.)

The direct-coupled amplifier is not very efficient and the losses increase as the number of stages increase. Because of the disadvantages, direct coupling is not used very often.

RC Coupling

The most commonly used coupling in amplifiers is RC coupling. An RC-coupling network is shown in figure 1-10.

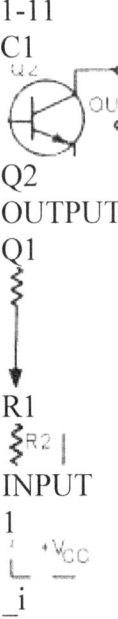

Figure 1-10.—RC-coupled transistor amplifier.

The network of Rl, R2, and CI enclosed in the dashed lines of the figure is the coupling network. You may notice that the circuitry for Q1 and Q2 is incomplete. That is intentional so that you can concentrate on the coupling network.

Rl acts as a load resistor for Q1 (the first stage) and develops the output signal of that stage. Do you remember how a capacitor reacts to ac and dc? The capacitor, CI, "blocks" the dc of Ql's collector, but "passes" the ac output signal. R2 develops this passed, or coupled, signal as the input signal to Q2 (the second stage). This arrangement allows the coupling of the signal while it isolates the biasing of each stage. This solves many of the problems associated with direct coupling.

RC coupling does have a few disadvantages. The resistors use dc power and so the amplifier has low efficiency. The capacitor tends to limit the low-frequency response of the amplifier and the amplifying device itself limits the high-frequency response. For audio amplifiers this is usually not a problem; techniques for overcoming these frequency limitations will be covered later in this module.

Before you move on to the next type of coupling, consider the capacitor in the RC coupling. You probably remember that capacitive reactance (X_c) is determined by the following formula:

$$X_c = \frac{1}{2\pi fC}$$

This explains why the low frequencies are limited by the capacitor. As frequency decreases, X_c increases. This causes more of the signal to be "lost" in the capacitor.

The formula for X_c also shows that the value of capacitance (C) should be relatively high so that capacitive reactance (X_c) can be kept as low as possible. So, when a capacitor is used as a coupling element, the capacitance should be relatively high so that it will couple the entire signal well and not reduce or distort the signal.

Impedance Coupling

Impedance coupling is very similar to RC coupling. The difference is the use of an impedance device (a coil) to replace the load resistor of the first stage.

Figure 1-11 shows an impedance-coupling network between two stages of amplification. LI is the load for Q1 and develops the output signal of the first stage. Since the d.c. resistance of a coil is low, the efficiency of the amplifier stage is increased. The amount of signal developed in the output of the stage depends on the inductive reactance of LI. Remember the formula for inductive reactance:

$$X_L = 2\pi fL$$

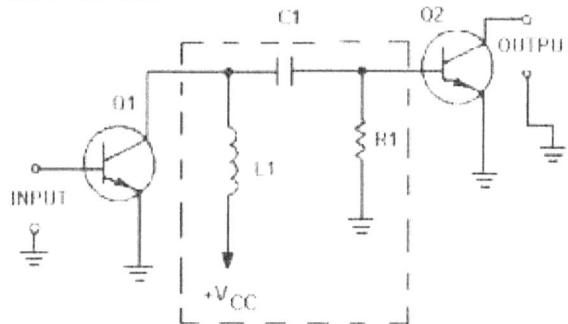

Figure 1-11.—Impedance-coupled transistor amplifier.

The formula shows that for inductive reactance to be large, either inductance or frequency or both must be high. Therefore, load inductors should have relatively large amounts of inductance and are most effective at high frequencies. This explains why impedance coupling is usually not used for audio amplifiers.

The rest of the coupling network (CI and Rl) functions just as their counterparts (CI and R2) in the RC-coupling network. CI couples the signal between stages while blocking the d.c. and Rl develops the input signal to the second stage (Q2).

Transformer Coupling

Figure 1-12 shows a transformer-coupling network between two stages of amplification. The transformer action of Tl couples the signal from the first stage to the second stage. In figure 1-12, the primary of Tl acts as the load for the first stage (Ql) and the secondary of Tl acts as the developing impedance for the second stage (Q2). No capacitor is needed because transformer action couples the signal between the primary and secondary of Tl.

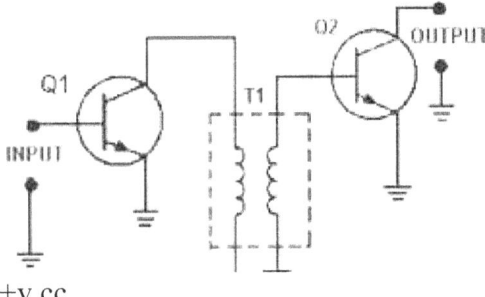

+v cc

Figure 1-12.—Transformer-coupled transistor amplifier.

The inductors that make up the primary and secondary of the transformer have very little dc resistance, so the efficiency of the amplifiers is very high. Transformer coupling is very often used for the final output (between the final amplifier stage and the output device) because of the impedance-matching qualities of the transformer. The frequency response of transformer-coupled amplifiers is limited by the inductive reactance of the transformer just as it was limited in impedance coupling.

Q-12. What is the purpose of an amplifier-coupling network?

Q-13. What are four methods of coupling amplifier stages?

Q-14. What is the most common form of coupling?

Q-15. What type coupling is usually used to couple the output from a power amplifier?

Q-16. What type coupling would be most useful for an audio amplifier between the first and second stages?

Q-17. What type of coupling is most effective at high frequencies?

IMPEDANCE CONSIDERATIONS FOR AMPLIFIERS

It has been mentioned that efficiency and impedance are important in amplifiers. The reasons for this may not be too clear. You have been shown that any amplifier is a current-control device. Now there are two other principles you should try to keep in mind. First, there is no such thing as "something for nothing" in electronics. That means every time you do something to a signal it costs something. It might mean a loss in fidelity to get high power. Some other compromise might also be made when a circuit is designed. Regardless of the compromise, every stage will require and use power. This brings up the second principle-do things as efficiently as possible. The improvement and design of electronic circuits is an attempt to do things as cheaply as possible, in terms of power, when all the other requirements (fidelity, power output, frequency range, etc.) have been met.

This brings us to efficiency. The most efficient device is the one that does the job with the least loss of power. One of the largest losses of power is caused by impedance differences between the output of one circuit and the input of the next circuit. Perhaps the best way to think of an impedance difference (mismatch) between circuits is to think of different-sized water pipes. If you try to connect a one-inch water pipe to a two-inch water pipe without an adapter you will lose water. You must use an adapter. A impedance-matching device is like that adapter. It allows the connection of two devices with different impedances without the loss of power.

Figure 1-13 shows two circuits connected together. Circuit number 1 can be considered as an a.c. source (E s) whose output impedance is represented by a resistor (Rl). It can be considered as an a.c. source because the output signal is an a.c. voltage and comes from circuit number 1 through the output impedance. The input impedance of circuit number 2 is represented by a resistor in series with the source. The resistance is shown as variable to show what will happen as the input impedance of circuit number 2 is changed.

Figure 1-13.—Effect of impedance matching in the coupling of two circuits.

The chart below the circuit shows the effect of a change in the input impedance of circuit number 2 (R2) on current (I), signal voltage developed at the input of circuit number 2 (E R2), the power at the output of circuit number 1 (P R i), and the power at the input to circuit number 2 (P R 2).

Two important facts are brought out in this chart. First, the power at the input to circuit number 2 is greatest when the impedances are equal (matched). The power is also equal at the output of circuit number 1 and the input of circuit number 2 when the impedance is matched. The second fact is that the largest voltage signal is developed at the input to circuit number 2 when its input impedance is much larger than the output impedance of circuit number 1. However, the power at the input of circuit number 2

is very low under these conditions. So you must decide what conditions you want in coupling two circuits together and select the components appropriately.

Two important points to remember about impedance matching are as follows. (1) Maximum power transfer requires matched impedance. (2) To get maximum voltage at the input of a circuit requires an intentional impedance mismatch with the circuit that is providing the input signal.

Impedance Characteristics of Amplifier Configurations

Now that you have seen the importance of impedance matching the stages in an electronic device, you may wonder what impedance characteristics an amplifier has. The input and output impedances of a transistor amplifier depend upon the configuration of the transistor. In Module 7, Introduction to Solid-State Devices and Power Supplies, you were introduced to the three transistor configurations; the common emitter, the common base, and the common collector. Examples of these configurations and their impedance characteristics are shown in figure 1-14.

MEDIUMINPUTZ 500ft-1500ft
~.'>CC f MEDIUM OUTPUT
f JOL-ft -FOtft
I

I
COMMON EMITTER

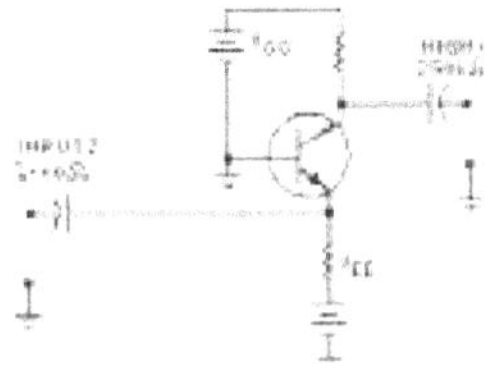

HIGH OUTPUTS 2F0kft - FFOkft
LOW INPUT Z 50ft-&0ft
COMMON BASE
HIGH INPUT 2
Elift - FFOfcft
"CO
H | I | -=l-
I
LOW OUTPUT 2 FOft- 1500ft
YEE
Hili-
He
I
COMMON COLLECTOR

Figure 1-14.—Transistor amplifier configurations and their impedance characteristics.

NOTE: Only approximate impedance values are shown. This is because the exact impedance values will vary from circuit to circuit. The impedance of any particular circuit depends upon the device (transistor) and the other circuit components. The value of impedance can be computed by dividing the signal voltage by the signal current. Therefore:

Input Signal Impedance
Input Signal Voltage
Input Signal Cunent
and
Output Signal Impedance
Output Signal Voltage
Output Signal Current

The common-emitter configuration provides a medium input impedance and a medium output impedance. The common-base configuration provides a low input impedance and a high output impedance. The common-collector configuration provides a high input impedance and a low output impedance. The common-collector configuration is often used to provide impedance matching between a high output impedance and a low input impedance.

If the amplifier stage is transformer coupled, the turns ratio of the transformer can be selected to provide impedance matching. In NEETS Module 2, Introduction to Alternating Current and Transformers, you were shown the relationship between the turns ratio and the impedance ratio in a transformer. The relationship is expressed in the following formula:

As you can see, impedance matching between stages can be accomplished by a combination of the amplifier configuration and the components used in the amplifier circuit.

Q-18. What impedance relationship between the output of one circuit and the input of another circuit will provide the maximum power transfer?

Q-19. If maximum current is desired at the input to a circuit, should the input impedance of that circuit be lower than, equal to, or higher than the output impedance of the previous stage?

Q-20. What are the input- and output-impedance characteristics of the three transistor configurations?

Q-21. What transistor circuit configuration should be used to match a high output impedance to a low input impedance?

Q-22. What type of coupling is most useful for impedance matching?

V/here:

N P

Number of turn5 in the primary Number of turns in the secondary Impedance of the primary Impedance of the secondary

AMPLIFIER FEEDBACK

Perhaps you have been around a public address system when a squeal or high-pitched noise has come from the speaker. Someone will turn down the volume and the noise will stop. That noise is an indication that the amplifier (at least one stage of amplification) has begun oscillating. Oscillation is covered in detail in NEETS Module 9, Introduction to Wave-Generation and Wave-Shaping Circuits. For now, you need only realize that the oscillation is caused by a small part of the signal from the amplifier output being sent back to the input of the amplifier. This signal is amplified and again sent back to the input where it is amplified again. This process continues and the result is a loud noise out of the speaker. The process of sending part of the output signal of an amplifier back to the input of the amplifier is called FEEDBACK.

There are two types of feedback in amplifiers. They are POSITIVE FEEDBACK, also called REGENERATIVE FEEDBACK, and NEGATIVE FEEDBACK, also called DEGENERATIVE FEEDBACK. The difference between these two types is whether the feedback signal is in phase or out of phase with the input signal.

Positive feedback occurs when the feedback signal is in phase with the input signal. Figure 1-15 shows a block diagram of an amplifier with positive feedback. Notice that the feedback signal is in phase with the input signal. This means that the feedback signal will add to or "regenerate" the input signal. The result is a larger amplitude output signal than would occur without the feedback. This type of feedback is what causes the public address system to squeal as described above.

FEEDBACK SIGNAL
FEEDBACK NETWORK
^7
INPUT SIGNAL

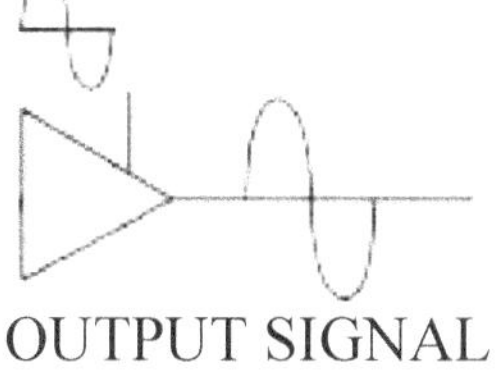

OUTPUT SIGNAL

Figure 1-15.—Positive feedback in an amplifier.

Figure 1-16 is a block diagram of an amplifier with negative feedback. In this case, the feedback signal is out of phase with the input signal. This means that the feedback signal will subtract from or "degenerate" the input signal. This results in a lower amplitude output signal than would occur without the feedback.

FEEDBACK SIGNAL
FEEDBACK NETWORK

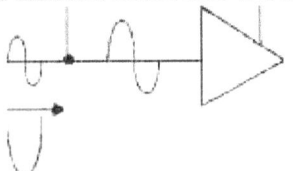

INPUT SIGNAL
OUTPUT SIGNAL

Figure 1-16.—Negative feedback in an amplifier.

Sometimes feedback that is not desired occurs in an amplifier. This happens at high frequencies and limits the high-frequency response of an amplifier. Unwanted feedback also occurs as the result of some circuit components used in the biasing or coupling network. The usual solution to unwanted feedback is a feedback network of the opposite type. For example, a positive feedback network would counteract unwanted, negative feedback.

Feedback is also used to get the ideal input signal. Normally, the maximum output signal is desired from an amplifier. The amount of the output signal from an amplifier is dependent on the amount of the input signal. However, if the input signal is too large, the amplifying device will be saturated and/or cut off during part of the input signal. This causes the output signal to be distorted and reduces the fidelity of the amplifier. Amplifiers must provide the proper balance of gain and fidelity.

Figure 1-17 shows the way in which feedback can be used to provide the maximum output signal without a loss in fidelity. In view A, an amplifier has good fidelity, but less gain than it could have. By adding some positive feedback, as in view B, the gain of the stage is increased. In view C, an amplifier has so much gain and such a large input signal that the output signal is distorted. This distortion is caused by the amplifying device becoming saturated and cutoff. By adding a negative feedback system, as in view D, the gain of the stage is decreased and the fidelity of the output signal improved.

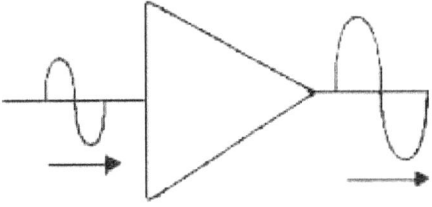

(A)

Figure 1-17A.-Feedback uses in amplifiers.

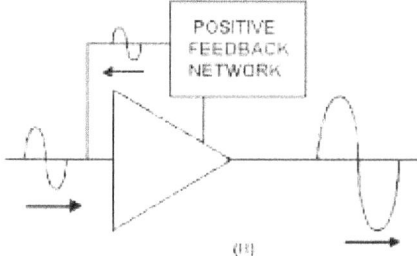

Figure 1-17B.—Feedback uses in amplifiers.

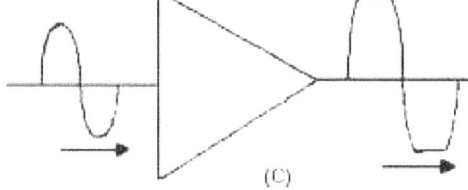

Figure 1-17C.—Feedback uses in amplifiers.

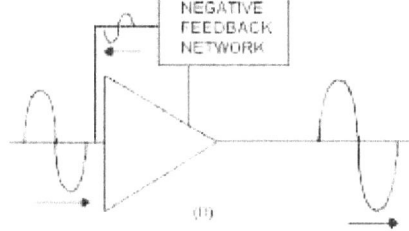

Figure 1-17D.—Feedback uses in amplifiers.

Positive and negative feedback are accomplished in many ways, depending on the reasons requiring the feedback. A few of the effects and methods of accomplishing feedback are presented next.

Positive Feedback

As you have seen, positive feedback is accomplished by adding part of the output signal in phase with the input signal. In a common-base transistor amplifier, it is fairly simple to provide positive feedback. Since the input and output signals are in phase, you need only couple part of the output signal back to the input. This is shown in figure 1-18.

The feedback network in this amplifier is made up of R2 and C2. The value of C2 should be large so that the capacitive reactance (X_c) will be low and the capacitor will couple the signal easily. (This is also the case with the input and output coupling capacitors CI and C3.) The resistive value of R2 should be large to limit the amount of feedback signal and to ensure that the majority of the output signal goes on to the next stage through C3.

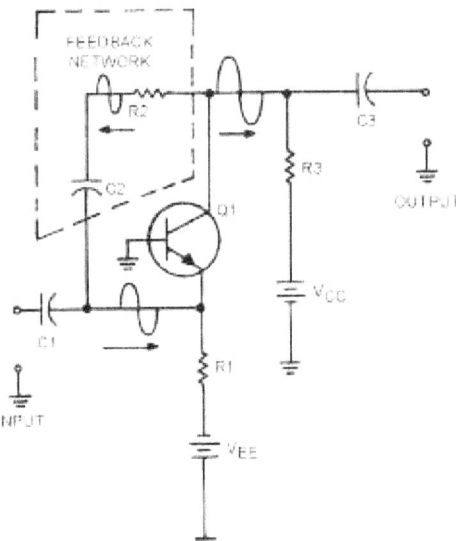

Figure 1-18.—Positive feedback in a transistor amplifier.

A more common configuration for transistor amplifiers is the common-emitter configuration. Positive feedback is a little more difficult with this configuration because the input and output signals are 180° out of phase. Positive feedback can be accomplished by feeding a portion of the output signal of the second stage back to the input of the first stage. This arrangement is shown in figure 1-19.

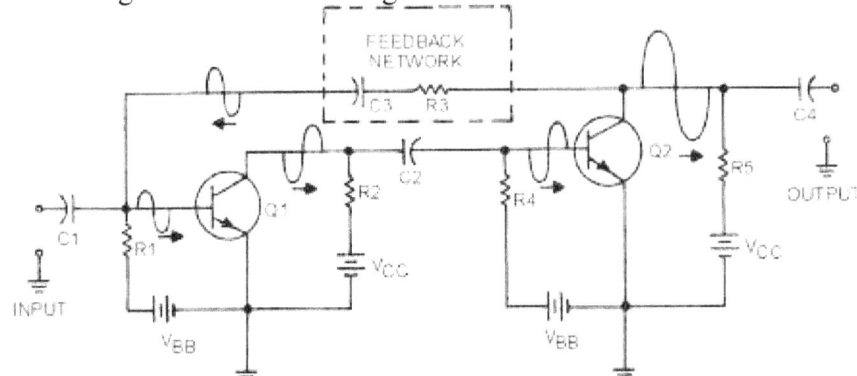

Figure 1-19.—Positive feedback in two stages of transistor amplification.

The figure shows that each stage of amplification has a 180° phase shift. This means that the output signal of Q2 will be in phase with the input signal to Q1. A portion of the output signal of Q2 is coupled back to the input of Q1 through the feedback network of C3 and R3. R3 should have a large resistance to limit the amount of signal through the feedback network. C3 should have a large capacitance so the capacitive reactance is low and the capacitor will couple the signal easily.

Sometimes positive feedback is used to eliminate the effects of negative feedback that are caused by circuit components. One way in which a circuit component can cause negative feedback is shown in figure 1-20.

In view (A) a common-emitter transistor amplifier is shown. An emitter resistor (R2) has been placed in this circuit to provide proper biasing and temperature stability. An undesired effect of this resistor is the development of a signal at the emitter in phase with the input signal on the base. This signal is caused by the changing current through the emitter resistor (R2) as the current through the transistor changes. You might think that this signal on the emitter is a form of positive feedback since it is in phase with the input signal. But the emitter signal is really negative feedback. Current through the transistor is controlled by the base-to-emitter bias. If both the base and emitter become more positive by the same amount at the same time, current will not increase. It is the difference between the base and emitter voltages that controls the current flow through the transistor.

To eliminate this negative feedback caused by the emitter resistor, some way must be found to remove the signal from the emitter. If the signal could be coupled to ground (decoupled) the emitter of the transistor would be unaffected. That is exactly what is done. A DECOUPLING CAPACITOR (C3 in view B) is placed between the emitter of Q1 and ground (across the emitter resistor). This capacitor should have a high capacitance so that it will pass the signal to ground easily. The decoupling capacitor (C3) should have the same qualities as the coupling capacitors (C1 and C2) of the circuit. Decoupling capacitors are also called bypass capacitors.

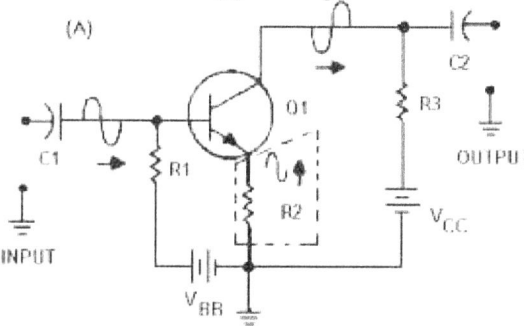

Figure 1-20A.—Decoupling (bypass) capacitor in a transistor amplifier.

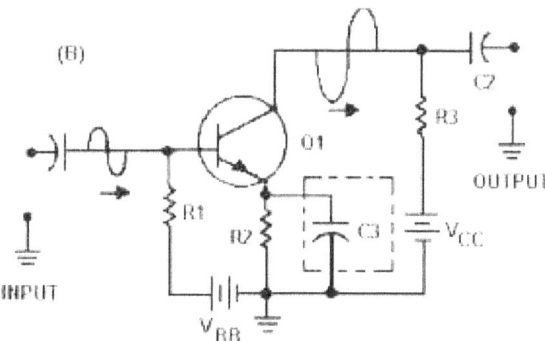

Figure 1-20B.—Decoupling (bypass) capacitor in a transistor amplifier.

Regardless of the method used to provide positive feedback in a circuit, the purpose is to increase the output signal amplitude.

Negative Feedback

Negative feedback is accomplished by adding part of the output signal out of phase with the input signal. You have seen that an emitter resistor in a common-emitter transistor amplifier will develop a negative feedback signal. Other methods of providing negative feedback are similar to those methods used to provide positive feedback. The phase relationship of the feedback signal and the input signal is the only difference.

Figure 1-21 shows negative feedback in a common-emitter transistor amplifier. The feedback network of C2 and R2 couples part of the output signal of Q1 back to the input. Since the output signal is 180° out of phase with the input signal, this causes negative feedback.

FEEDBACK NETWORK
R2
C1
INPUT 
Q1
R1
C3 R3
OUTPUT
V CC

Figure 1-21.—Negative feedback in a transistor amplifier.

Negative feedback is used to improve fidelity of an amplifier by limiting the input signal. Negative feedback can also be used to increase the frequency response of an amplifier. The gain of an amplifier decreases when the limit of its frequency response is reached. When negative feedback is used, the feedback signal decreases as the output signal decreases. At the limits of frequency response of the amplifier the smaller feedback signal means that the effective gain (gain with feedback) is increased. This will improve the frequency response of the amplifier.

Q-23. What is feedback?

Q-24. What are the two types of feedback?

Q-25. What type feedback provides increased amplitude output signals?

Q-26. What type feedback provides the best fidelity?

Q-27. If the feedback signal is out ofphase with the input signal, what type feedback is provided?

Q-28. What type feedback is provided by an unbypassed emitter resistor in a common-emitter transistor amplifier?

AUDIO AMPLIFIERS

An audio amplifier has been described as an amplifier with a frequency response from 15 Hz to 20 kHz. The frequency response of an amplifier can be shown graphically with a frequency response curve. Figure 1-22 is the ideal frequency response curve for an audio amplifier. This curve is practically "flat" from 15 Hz to 20 kHz. This means that the gain of the amplifier is equal between 15 Hz and 20 kHz. Above 20 kHz or below 15 Hz the gain decreases or "drops off quite rapidly. The frequency response of an amplifier is determined by the components in the circuit.

GAIN
0 10 20
I
5k 10k

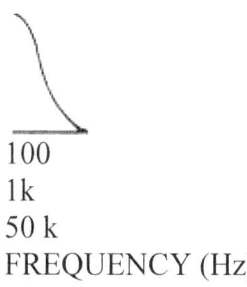

100
1k
50 k
FREQUENCY (Hz)

Figure 1-22-Ideal frequency response curve for an audio amplifier.

The difference between an audio amplifier and other amplifiers is the frequency response of the amplifier. In the next chapter of this module you will be shown the techniques and components used to change and extend the frequency response of an amplifier.

The transistor itself will respond quite well to the audio frequency range. No special components are needed to extend or modify the frequency response.

You have already been shown the purpose of all the components in a transistor audio amplifier. In this portion of the chapter, schematic diagrams of several audio amplifiers will be shown and the functions of each of the components will be discussed.

SINGLE-STAGE AUDIO AMPLIFIERS

The first single-stage audio amplifier is shown in figure 1-23. This circuit is a class A, common-emitter, RC-coupled, transistor, audio amplifier. CI is a coupling capacitor that couples the input signal to the base of Ql. Rl is used to develop the input signal and provide bias for the base of Ql. R2 is used to bias the emitter and provide temperature stability for Ql. C2 is used to provide decoupling (positive feedback) of the signal that would be developed by R2. R3 is the collector load for Ql and develops the output signal. C3 is a coupling capacitor that couples the output signal to the next stage. V C c represents the collector-supply voltage. Since the transistor is a common-emitter configuration, it provides voltage amplification. The input and output signals are 180° out of phase. The input and output impedance are both medium.

INPUT r\' SIGNAL U

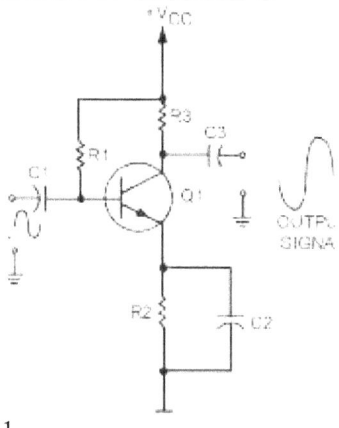

1
OUTPUT SIGNAL

Figure 1-23.—Transistor audio amplifier.

There is nothing new presented in this circuit. You should understand all of the functions of the components in this circuit. If you do not, look back at the various sections presented earlier in this chapter.

The second single-stage audio amplifier is shown in figure 1-24. This circuit is a class A, common-source, RC-coupled, FET, audio amplifier. CI is a coupling capacitor which couples the input signal to the gate of Ql. Rl is used to develop the input signal for the gate of Ql. R2 is used to bias the source of Ql. C2 is used to decouple the signal developed by R2 (and keep it from affecting the source of Ql). R3 is the drain load for Ql and develops the output signal. C3 couples the output signal to the next stage. V DD is the supply voltage for the drain of Ql. Since this is a common-source configuration, the input and output signals are 180° out of phase.

+V
INPUT SIGNAL
1
DD

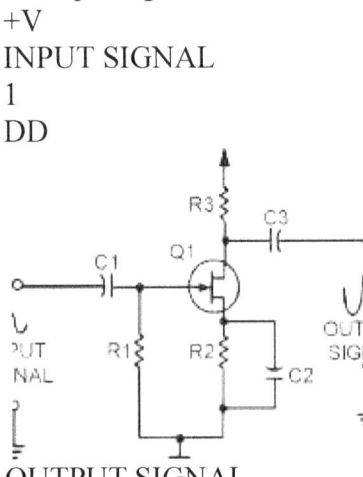

OUTPUT SIGNAL

Figure 1-24.—FET audio amplifier.

If you do not remember how a FET works, refer to NEETS Module 7 Introduction to Solid-State Devices and Power Supplies.

The third single-stage audio amplifier is shown in figure 1-25. This is a class A, common-emitter, transformer-coupled, transistor, audio amplifier. The output device (speaker) is shown connected to the secondary winding of the transformer. CI is a coupling capacitor which couples the input signal to the base of Ql. Rl develops the input signal. R2 is used to bias the emitter of Ql and provides temperature stability. C2 is a decoupling capacitor for R2. R3 is used to bias the base of Ql. The primary of Tl is the collector load for Ql and develops the output signal. Tl couples the output signal to the speaker and provides impedance matching between the output impedance of the transistor (medium) and the impedance of the speaker (low).

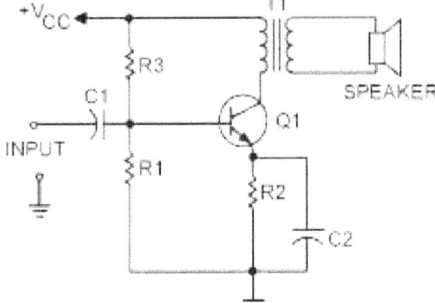

Figure 1-25.—Single-stage audio amplifier.

PHASE SPLITTERS

Sometimes it is necessary to provide two signals that are equal in amplitude but 180° out of phase with each other. (You will see one use of these two signals a little later in this chapter.) The two signals can be provided from a single input signal by the use of a PHASE SPLITTER. A phase splitter is a device that produces two signals that differ in phase from each other from a single input signal. Figure 1-26 is a block diagram of a phase splitter.

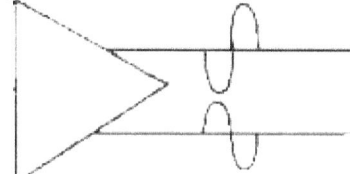

Figure 1-26.—Block diagram of a phase splitter.

One way in which a phase splitter can be made is to use a center-tapped transformer. As you may remember from your study of transformers, when the transformer secondary winding is center-tapped, two equal amplitude signals are produced. These signals will be 180° out of phase with each other. So a transformer with a center-tapped secondary fulfills the definition of a phase splitter.

A transistor amplifier can be configured to act as a phase splitter. One method of doing this is shown in figure 1-27.

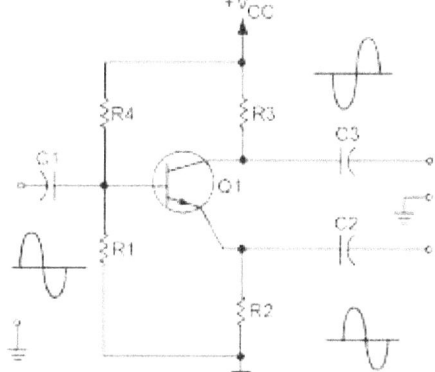

Figure 1-27.—Single-stage transistor phase splitter.

CI is the input signal coupling capacitor and couples the input signal to the base of Q1. R1 develops the input signal. R2 and R3 develop the output signals. R2 and R3 are equal resistances to provide equal amplitude output signals. C2 and C3 couple the output signals to the next stage. R4 is used to provide proper bias for the base of Q1.

This phase splitter is actually a single transistor combining the qualities of the common-emitter and common-collector configurations. The output signals are equal in amplitude of the input signal, but are 180° out of phase from each other.

If the output signals must be larger in amplitude than the input signal, a circuit such as that shown in figure 1-28 will be used.

Figure 1-28 shows a two-stage phase splitter. CI couples the input signal to the base of Q1. R1 develops the input signal and provides bias for the base of Q1. R2 provides bias and temperature stability for Q1. C2 decouples signals from the emitter of Q1. R3 develops the output signal of Q1. Since Q1 is configured as a common-emitter amplifier, the output signal of Q1 is 180° out of phase with the input signal and larger in amplitude. C3 couples this output signal to the next stage through R4. R4 allows only a small portion of this output signal to be applied to the base of Q2. R5 develops the input signal and provides bias for the base of Q2. R6 is used for bias and temperature stability for Q2. C4 decouples signals from the emitter of Q2. R7 develops the output signal from Q2. Q2 is configured as a common-emitter amplifier, so the output signal is 180° out of phase with the input signal to Q2 (output signal from Ql). The input signal to Q2 is 180° out of phase with the original input signal, so the output from Q2 is in phase with the original input signal. C5 couples this output signal to the next stage. So the circuitry shown provides two output signals that are 180° out of phase with each other. The output signals are equal in amplitude with each other but larger than the input signal.

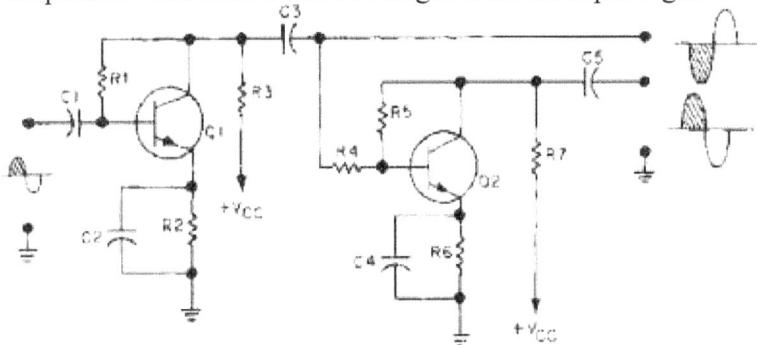

Figure 1-28.—Two-stage transistor phase splitter.

Q-29. What is a phase splitter?

PUSH-PULL AMPLIFIERS

One use of phase splitters is to provide input signals to a single-stage amplifier that uses two transistors. These transistors are configured in such a way that the two outputs, 180° out of phase with each other, combine. This allows more gain than one transistor could supply by itself. This "push-pull" amplifier is used where high power output and good fidelity are needed: receiver output stages, public address amplifiers, and AM modulators, for example.

The circuit shown in figure 1-29 is a class A transistor push-pull amplifier, but class AB or class B operations can be used. Class operations were discussed in an earlier topic. The phase splitter for this amplifier is the transformer T1, although one of the phase splitters shown earlier in this topic could be used. Rl provides the proper bias for Ql and Q2. The tapped secondary of Tl develops the two input signals for the bases of Ql and Q2. Half of the original input signal will be amplified by Q-l, the other half by Q-2. T2 combines (couples) the amplified output signal to the speaker and provides impedance matching.

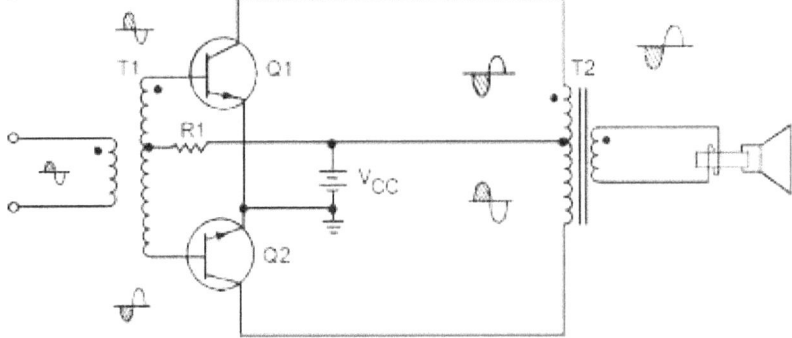

Figure 1-29.—Class A transistor push-pull amplifier.

Q-30. What is one use for a splitter?
Q-31. What is a common use for a push-pull amplifier?
Q-32. What is the advantage of a push-pull amplifier?
Q-33. What class of operation can be used with a push-pull amplifier to provide

good fidelity output signals?

SUMMARY

This chapter has presented some general information that applies to all amplifiers, as well as some specific information about transistor and audio amplifiers. All of this information will be useful to you in the next chapter of this module and in your future studies of electronics.

An AMPLIFIER is a device that enables an input signal to control an output signal. The output signal will have some (or all) of the characteristics of the input signal but will generally be larger than the input signal in terms of voltage, current, or power. A basic line diagram of an amplifier is shown below.

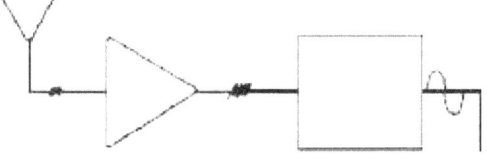

AMPLIFIER (RADIO FREQUENCY)
DETECTOR (SEPARATES AUDIO FROM RADIO FREQUENCY)

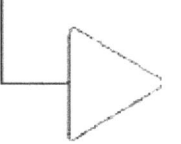

AMPLIFIER (AUDIO FREQUENCY)

Amplifiers are classified by FUNCTION and FREQUENCY RESPONSE. Function refers to an amplifier being a VOLTAGE AMPLIFIER or a POWER AMPLIFIER. Voltage amplifiers provide voltage amplification and power amplifiers provide power amplification. The frequency response of an amplifier can be described by classifying the amplifier as an AUDIO AMPLIFIER, RF AMPLIFIER, or VIDEO (WIDE-BAND) AMPLIFIER. Audio amplifiers have frequency response in the range of 15 Hz to 20 kHz. An rf amplifier has a frequency response in the range of 10 kHz to 100,000 MHz. A video (wide-band) amplifier has a frequency response of 10 Hz to 6 MHz.

INPUT SIGNAL 200 mV PEAK-TO-PEAK
OUTPUT SIGNAL 4 V PEAK-TO-PEAK

INPUT SIGNAL OUTPUT SIGNAL
10 V PEAK-TO-PEAK 5 V PEAK-TO-PEAK
100 mA(1W OF POWER) 2 A (10 W OF POWER)

The CLASS OF OPERATION of a transistor amplifier is determined by the percent of time that current flows through the transistor in relation to the input signal.

In CLASS A OPERATION, transistor current flows for 100% (360°) of the input signal. Class A operation is the least efficient class of operation, but provides the best fidelity.

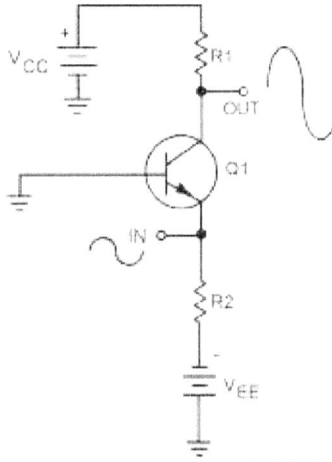

In CLASS AB OPERATION, transistor current flows for more than 50% but less than 100% of the input signal.

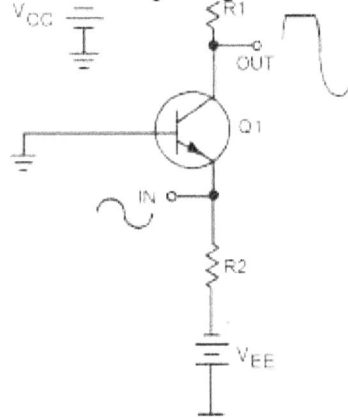

In CLASS B OPERATION, transistor current flows for 50% of the input signal.

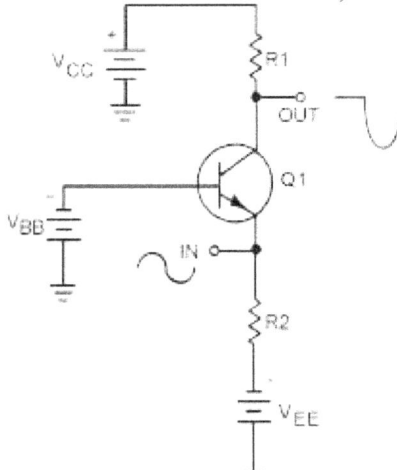

In CLASS C OPERATION, transistor current flows for less than 50% of the input signal. Class C operation is the most efficient class of operation.

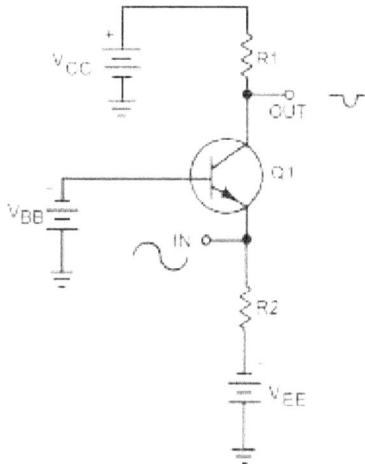

COUPLING is used to transfer a signal from one stage to another.

DIRECT COUPLING is the connection of the output of one stage directly to the input of the next stage. This method is not used very often due to the complex power supply requirements and impedance-matching problems.

V
R1«
o— 4-
INPUT
1
BB
Q1
7V

R4 4—VW-
■±r 6V
Q2

R2i
R5 ■VW-
13V
♦ o
OUTPUT
R
R6
10V
15V
.V
CC

RC COUPLING is the most common method of coupling and uses a coupling capacitor and signal-developing resistors.

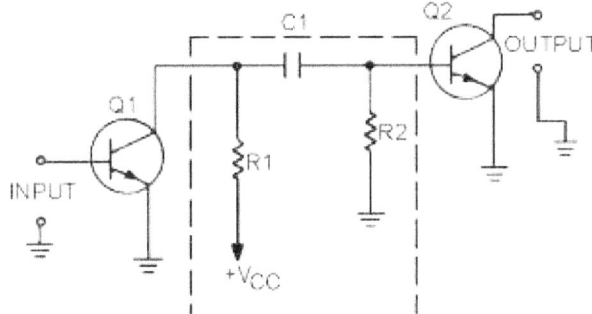

IMPEDANCE COUPLING uses a coil as a load for the first stage but otherwise functions just as RC coupling. Impedance coupling is used at high frequencies.

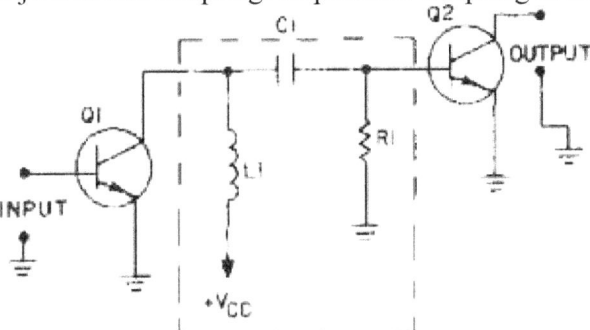

TRANSFORMER COUPLING uses a transformer to couple the signal from one stage to the next. Transformer coupling is very efficient and the transformer can aid in impedance matching.

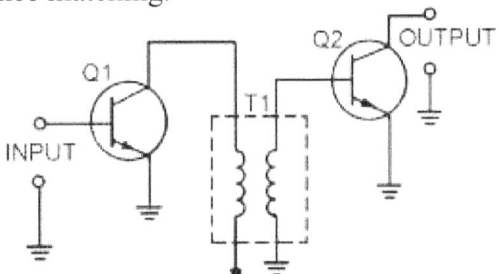

+ vcc

MAXIMUM POWER TRANSFER occurs between two circuits when the output impedance of the first circuit matches the input impedance of the second circuit.

A MAXIMUM VOLTAGE INPUT SIGNAL is present when the input impedance of the second circuit is larger than the output impedance of the first circuit (mismatched).

The COMMON-EMITTER configuration of a transistor amplifier has MEDIUM INPUT and MEDIUM OUTPUT IMPEDANCE.

MEDIUM INPUT 2 500ft-1S00ft
X
MEDIUM OUTPUT SOkft -SOkft
It
1

IHPUT
EE
Hili-
I
OUTPUT
COMMON EMITTER

The COMMON-BASE configuration of a transistor amplifier has LOW INPUT and HIGH OUTPUT IMPEDANCE.

CO
HIGH OUTPUTS ESOkft - 55iikft
LOW IHPUT 2 30ft-t0ft
It
1
OUTPUT
EE
IHPUT X
COMMON BASE

The COMMON-COLLECTOR (EMITTER FOLLOWER) configuration of a transistor amplifier as HIGH INPUT and LOW OUTPUT IMPEDANCE.

HIGH INPUTS 2\tSi. - 550kft

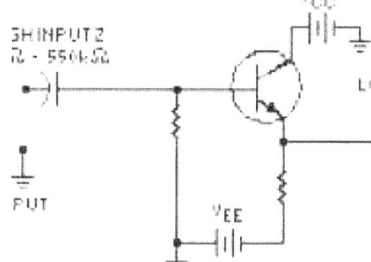

INPUT
LOW OUTPUTS 50 Si.- 1500ft
If
1
OUTPUT
COMMON COLLECTOR

FEEDBACK is the process of coupling a portion of the output signal back to the input of an amplifier.

POSITIVE (REGENERATIVE) FEEDBACK is provided when the feedback signal is in phase with the input signal. Positive feedback increases the gain of an amplifier.

FEEDBACK SIGNAL
FEEDBACK NETWORK
^7
INPUT SIGNAL

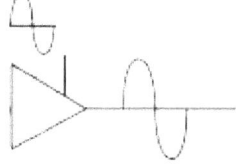

OUTPUT SIGNAL

NEGATIVE (DEGENERATIVE) FEEDBACK is provided when the feedback signal is 180° out of phase with the input signal. Negative feedback decreases the gain of an amplifier but improves fidelity and may increase the frequency response of the amplifier.

FEEDBACK SIGNAL
t/ 1
FEEDBACK NETWORK

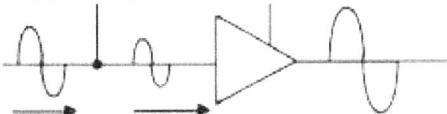

INPUT INPUT SIGNAL SIGNAL WITH FEEDBACK
OUTPUT SIGNAL

The IDEAL FREQUENCY RESPONSE of an audio amplifier is equal gain from 15 Hz to 20 kHz
with very low gain outside of those limits.

GAIN

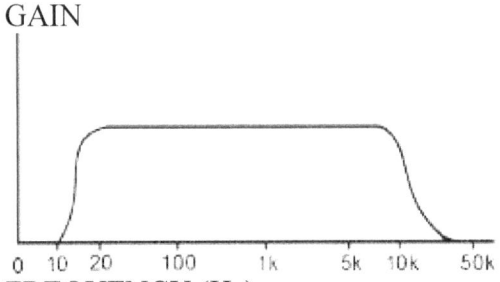

FREQUENCY (Hz)

A PHASE SPLITTER provides two output signals that are equal in amplitude but different in phase from a single input signal. Phase splitters are often used to provide input signals to a push-pull amplifier.

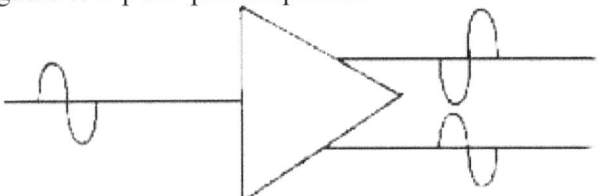

A PUSH-PULL AMPLIFIER uses two transistors whose output signals are added together to provide a larger gain (usually a power gain) than a single transistor could provide. Push-pull amplifiers can be operated class A, class AB or class B.

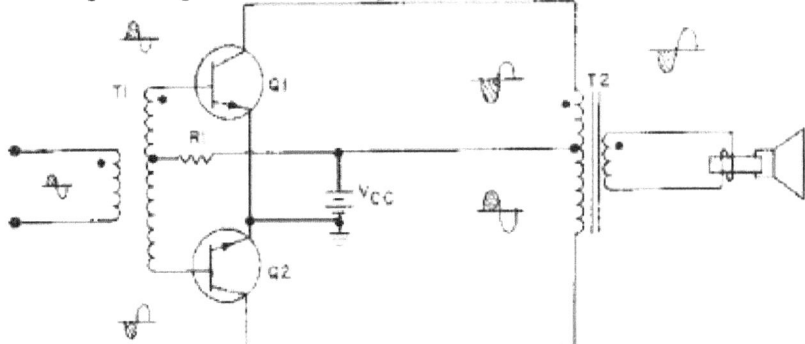

ANSWERS TO QUESTIONS Q1. THROUGH Q33.

A-1. Amplification is the control of an output signal by an input signal so that the output signal has some (or all) of the characteristics of the input signal. The output signal is generally larger than the input signal in terms of voltage, current, or power.

A-2. No, the input signal is unchanged, the output signal is controlled by the input signal but does not effect the actual input signal.

A-3. To amplify the input signal to a usable level.

A-4. By function and frequency response.

A-5. An audio power amplifier.

A-6. An rf voltage amplifier.

A-7. The amount of time (in relation to the input signal) in which current flows in the output circuit.

A-8. A, AB, B, C.

A-9. Class B operation.

A-10. The amplifier operates (and therefore uses power) for less time in class C than in class A.

A-1 I. Class A operation.

A-12. To transfer energy (a signal) from one stage to another.

A-13. Direct, RC, impedance, and transformer coupling.

A-14. RC coupling.

A-15. Transformer coupling.

A-16. RC coupling.

A-17. Impedance coupling.

A-18. Equal impedance.

A-19. Lower than.

A-20. Common emitter-medium input, medium output; common base-low input, high output; common collector-high input, low output.

A-21. Common collector.

A-22. Transformer coupling.

A-23. The process of coupling a portion of the output of a circuit back to the circuit input.

A-24. Positive and negative or regenerative and degenerative.

A-25. Positive (regenerative) feedback.

A-26. Negative (degenerative) feedback.

A-27. Negative (degenerative) feedback.

A-28. Negative (degenerative) feedback.

A-29. A device that provides two output signals that differ in phase from a single input signal.

A-30. A phase splitter is used to provide the input signals to a push-pull amplifier.

A-31. A push-pull amplifier is used when high power output and good fidelity are needed.

A-32. A push-pull amplifier provides more gain than a single transistor amplifier.

A-33. Class A, Class AB or Class B operation.

CHAPTER 2

VIDEO AND RF AMPLIFIERS

LEARNING OBJECTIVES

Upon completion of this chapter, you will be able to:

1. Define the term "bandwidth of an amplifier."
2. Determine the upper and lower frequency limits of an amplifier from a frequency-response curve.
3. List the factors that limit frequency response in an amplifier.
4. List two techniques used to increase the high-frequency response for a video amplifier.
5. State one technique used to increase the low-frequency response of a video amplifier.
6. Identify the purpose of various components on a schematic of a complete typical video amplifier circuit.
7. State the purpose of a frequency-determining network in an rf amplifier.
8. State one method by which an rf amplifier can be neutralized.
9. Identify the purpose of various components on a schematic of a complete typical rf amplifier.

INTRODUCTION

In this chapter you will be given information on the frequency response of amplifiers as well as specific information on video and rf amplifiers. For all practical purposes, all the general information you studied in chapter 1 about audio amplifiers will apply to the video and rf amplifiers which you are about to study.

You may be wondering why you need to learn about video and rf amplifiers. You need to understand these circuits because, as a technician, you will probably be involved in working on equipment in which these circuits are used. Many of the circuits shown in this and the next chapter are incomplete and would not be used in actual equipment. For example, the complete biasing network may not be shown. This is done so you can concentrate on the concepts being presented without being overwhelmed by an abundance of circuit elements. With this idea in mind, the information that is presented in this chapter is real, practical information about video and rf amplifiers. It is the sort of information that you will use in working with these circuits. Engineering information (such as design specifications) will not be presented because it is not needed to understand the concepts that a technician needs to perform the job of circuit analysis and repair. Before you are given the specific information on video and rf amplifiers, you may be wondering how these circuits are used.

Video amplifiers are used to amplify signals that represent video information. (That's where the term "video" comes from.) Video is the "picture" portion of a television signal. The "sound" portion is audio.

Although the Navy uses television in many ways, video signals are used for more than television. Radar systems (discussed later in this training series) use video signals and, therefore, video amplifiers. Video amplifiers are also used in video recorders and some communication and control devices. In addition to using video amplifiers, televisions use rf amplifiers. Many other devices also use rf amplifiers, such as radios, navigational devices, and communications systems. Almost any device that uses broadcast, or transmitted, information will use an rf amplifier.

As you should recall, rf amplifiers are used to amplify signals between 10 kilohertz (10 kHz) and 100,000 megahertz (100,000 MHz) (not this entire band of frequencies, but any band of frequencies within these limits). Therefore, any device that uses frequencies between 10 kilohertz and 100,000 megahertz will most likely use an rf amplifier.

Before you study the details of video and rf amplifiers, you need to learn a little more about the frequency response of an amplifier and frequency-response curves.

AMPLIFIER FREQUENCY RESPONSE

In chapter 1 of this module you were shown the frequency-response curve of an audio amplifier. Every amplifier has a frequency-response curve associated with it. Technicians use frequency-response curves because they provide a "picture" of the performance of an amplifier at various frequencies. You will probably never have to draw a frequency-response curve, but, in order to use one, you should know how a frequency-response curve is created. The amplifier for which the frequency-response curve is created is tested at various frequencies. At each frequency, the input signal is set to some predetermined level of voltage (or current). This same voltage (or current) level for all of the input signals is used to provide a standard input and to allow evaluation of the output of the circuit at each of the frequencies tested. For each of these frequencies, the output is measured and marked on a graph. The graph is marked "frequency" along the horizontal axis and "voltage" or "current" along the vertical axis. When points have been plotted for all of the frequencies tested, the points are connected to form the frequency-response curve. The shape of the curve represents the frequency response of the amplifier.

Some amplifiers should be "flat" across a band of frequencies. In other words, for every frequency within the band, the amplifier should have equal gain (equal response). For frequencies outside the band, the amplifier gain will be much lower.

For other amplifiers, the desired frequency response is different. For example, perhaps the amplifier should have high gain at two frequencies and low gain for all other frequencies. The frequency-response curve for this type of amplifier would show two "peaks." In other amplifiers the frequency-response curve will have one peak indicating high gain at one frequency and lower gain at all others.

Note the frequency-response curve shown in figure 2-1. This is the frequency-response curve for an audio amplifier as described in chapter 1. It is "flat" from 15 hertz (15 Hz) to 20 kilohertz (20 kHz).

0 10 50 100 1k 10k 100k 1M 10M FREQUENCY (Hz)

Figure 2-1.—Frequency response curve of audio amplifier.

Notice in the figure that the lower frequency limit is labeled f_1 and the upper frequency limit is labeled f_2. Note also the portion inside the frequency-response curve marked "BANDWIDTH." You may be wondering just what a "bandwidth" is.

BANDWIDTH OF AN AMPLIFIER

The bandwidth represents the amount or "width" of frequencies, or the "band of frequencies," that the amplifier is MOST effective in amplifying. However, the bandwidth is NOT the same as the band of frequencies that is amplified. The bandwidth (BW) of an amplifier is the difference between the frequency limits of the amplifier. For example, the band of frequencies for an amplifier may be from 10 kilohertz (10 kHz) to 30 kilohertz (30 kHz). In this case, the bandwidth would be 20 kilohertz (20 kHz). As another example, if an amplifier is designed to amplify frequencies between 15 hertz (15 Hz) and 20 kilohertz (20 kHz), the bandwidth will be equal to 20 kilohertz minus 15 hertz or 19,985 hertz (19,985 Hz). This is shown in figure 2-1.

Mathematically:

$BW = f_2 - f_1$

= 20 kHz - 15 Hz BW = 20,000 Hz - 15Hz BW= 19,985 Hz

You should notice on the figure that the frequency-response curve shows output voltage (or current) against frequency. The lower and upper frequency limits (f_1 and f_2) are also known as HALF-POWER POINTS. The half-power points are the points at which the output voltage (or current) is 70.7 percent of the maximum output voltage (or current). Any frequency that produces less than 70.7 percent of the maximum output voltage (or current) is outside the bandwidth and, in most cases, is not considered a useable output of the amplifier.

The reason these points are called "half-power points" is that the true output power will be half (50 percent) of the maximum true output power when the output voltage (or current) is 70.7 percent of the maximum output voltage (or current), as shown below. (All calculations are rounded off to two decimal places.)

As you learned in NEETS, Module 2, in an a.c. circuit true power is calculated using the resistance (R) of the circuit, NOT the impedance (Z). If the circuit produces a maximum output voltage of 10 volts across a 50-ohm load, then:

True Power = — R
True Power
50fi
True Power = watts
50
True Power =2 watts

When the output voltage drops to 70.7 percent of the maximum voltage of 10 volts, then:

True Power =
E 2
R
_ n (7.07V) 2 True Power =
SOQ
True Power = — watts 50
True Power =1 watts

As you can see, the true power is 50 percent (half) of the maximum true power when the output voltage is 70.7 percent of the maximum output voltage. If, instead, you are using the output current of the above circuit, the maximum current is

,2arnp
f—--4
I SOQ)

The calculations are:
True Power = I 2 R   True Power = (,2A) 2 (50Q)   True Power = (.04) (50) watts   True Power = 2 watts

At 70.7 percent of the output current (.14 A):
2-4
TruePower = PR   TruePower = (.14A) 2 (50Q)   TruePower = (.02 x50)watt5   TruePower = lwatt

On figure 2-1, the two points marked f_1 and f_2 will enable you to determine the frequency-response limits of the amplifier. In this case, the limits are 15 hertz (15 Hz) and 20 kilohertz (20 kHz). You should now see how a frequency-response curve can enable you to determine the frequency limits and the bandwidth of an amplifier.

READING AMPLIFIER FREQUENCY-RESPONSE CURVES

Figure 2-2 shows the frequency-response curves for four different amplifiers. View (A) is the same frequency-response curve as shown in figure 2-1. View (B) is the frequency-response curve of an amplifier that would also be classified as an audio amplifier, even though the curve is not "flat" from 15 hertz to 20 kilohertz and does not drop off sharply at the frequency limits. From the curve, you can see that the lower frequency limit of this amplifier (f1) is 100 hertz. The upper frequency limit (f2) is 10 kilohertz. Therefore, the bandwidth of this amplifier must be 10 kilohertz minus 100 hertz or 9900 hertz. Most amplifiers will have a frequency-response curve shaped like view (B) if nothing is done to modify the frequency-response characteristics of the circuit. (The factors that affect frequency response and the methods to modify the frequency response of an amplifier are covered a little later in this chapter.)

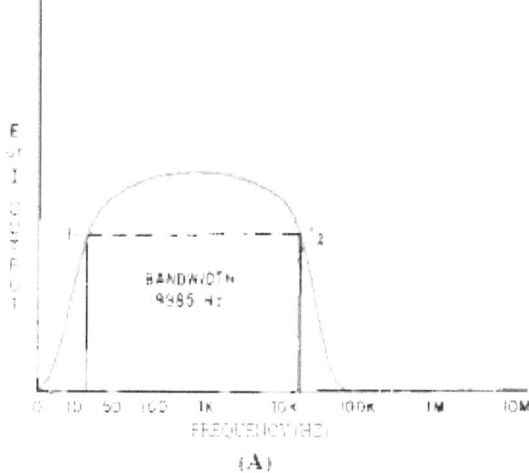

Figure 2-2A. — Frequency response curves.

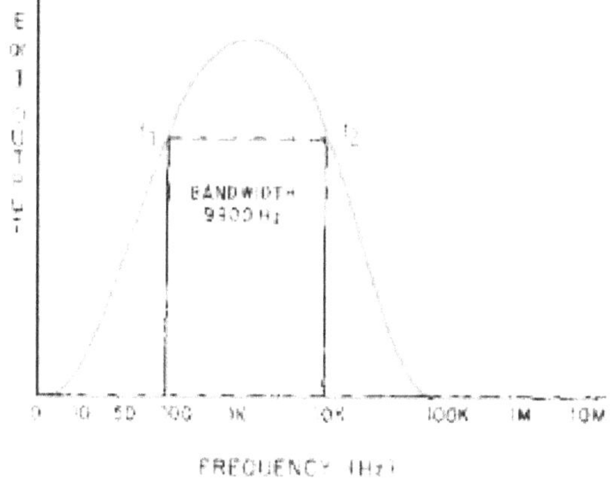

Figure 2-2B. — Frequency response curves.

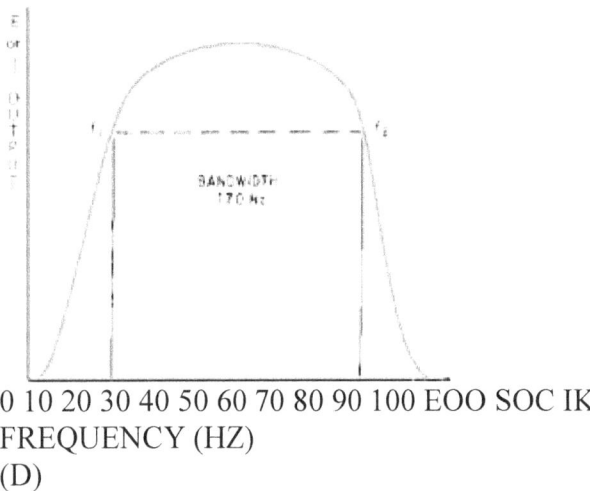

0 10 20 30 40 50 60 70 80 90 100 EOO SOC IK
FREQUENCY (HZ)
(D)

Figure 2-2C.—Frequency response curves.

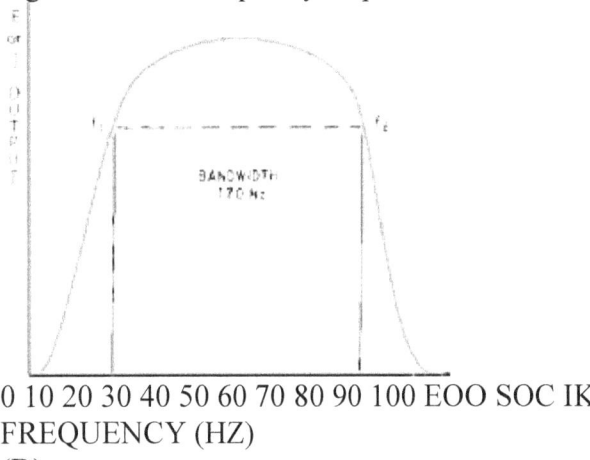

0 10 20 30 40 50 60 70 80 90 100 EOO SOC IK
FREQUENCY (HZ)
(D)

Figure 2-2D.—Frequency response curves.

Now look at view (C). This frequency-response curve is for an rf amplifier. The frequency limits of this amplifier are 100 kilohertz (fi) and 1 megahertz (f 2); therefore, the bandwidth of this amplifier is 900 kilohertz.

View (D) shows another audio amplifier. This time the frequency limits are 30 hertz (f\'7b) and 200 hertz (f 2). The bandwidth of this amplifier is only 170 hertz. The important thing to notice in view (D) is that the frequency scale is different from those used in other views. Any frequency scale can be used for a frequency-response curve. The scale used would be determined by what frequencies are most useful in presenting the frequency-response curve for a particular amplifier.

Q-l. What is the bandwidth of an amplifier?

Q-2. What are the upper and lower frequency limits of an amplifier?

Q-3. What are the upper and lower frequency limits and the bandwidth for the amplifiers that have frequency-response curves as shown in figure 2-3?

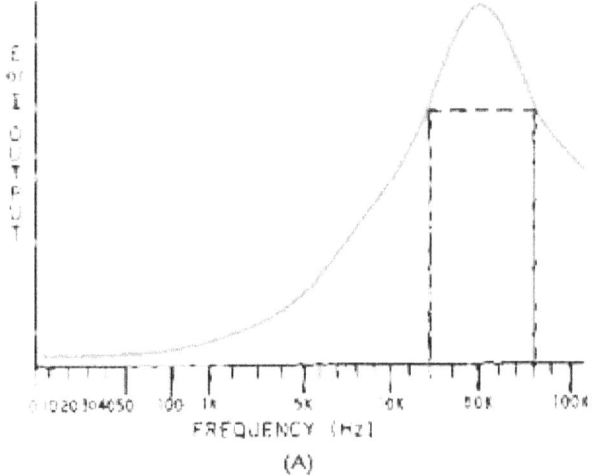

Figure 2-3A.—Frequency-response curves for Q3.

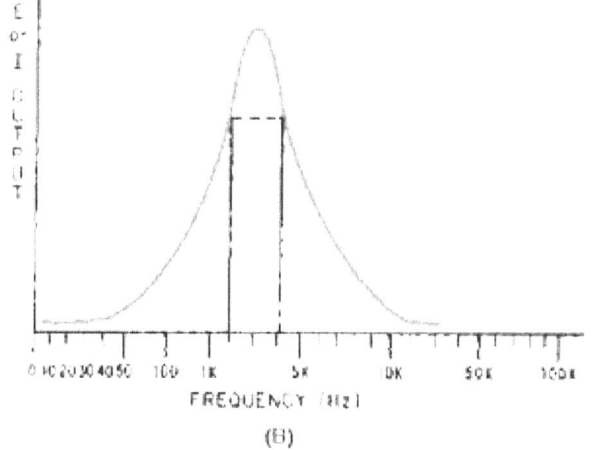

Figure 2-3B.—Frequency-response curves for Q3.

FACTORS AFFECTING FREQUENCY RESPONSE OF AN AMPLIFIER

In chapter 1 of this module, the fact was mentioned that an audio amplifier is limited in its frequency response. Now you will see why this is true.

You should recall that the frequency response of an a.c. circuit is limited by the reactive elements (capacitance and inductance) in the circuit. As you know, this is caused by the fact that the capacitive and inductive reactances vary with the frequency. In other words, the value of the reactance is determined, in part, by frequency. Remember the formulas:

L 2n£C X L = 2n£L

If you ignore the amplifying device (transistor, electron tube, etc.), and if the amplifier circuit is made up of resistors only, there should be no limits to the frequency response. In other words, a totally resistive circuit would have no frequency limits. However, there is no such thing as a totally resistive circuit because circuit components almost always have some reactance. In addition to the reactance of other components in the circuit, most amplifiers use RC coupling. This means that a capacitor is used to couple the signal in to and out of the circuit. There is also a certain amount of capacitance and inductance in the wiring of the circuit. The end result is that all circuits are reactive. To illustrate this point, figure 2-4 shows amplifier circuits with the capacitance and inductance of the wiring represented as "phantom" capacitors and inductors. The reactances of the capacitors (X c) and the inductors (X L) are shown as "phantom" variable resistors. View (A) shows the circuit with a low-frequency input signal, and view (B) shows the circuit with a high-frequency input signal.

'CI
Hi
LOW .-v£L, r -_
FREQUENCY ! ! ■ !
L2 "C3 , -^vJ- 1 j-«vfl--i
^1
CI ■ L.
LI
o
C3
R3
" =1= ?*3 X cr R2 4 ■
V
C2 =j= ^±a x c2 A.
MA)

Figure 2-4A.—Amplifiers showing reactive elements and reactance.

HIGH ^ t ,fl r
FREQUENCY ■ ; j !
ci
L1 j
C4
i i
i
.1.
CR
rO
R1
1.2
I -tvJ- i j---vfl--i
C3
'R3
C2;
=!= ^-1 X C2
r—i
* J
(A)

Figure 2-4B.—Amplifiers showing reactive elements and reactance.

The actual circuit components are: CI, C2, C3, Rl, R2, R3, and Ql. CI is used to couple the input signal. Rl develops the input signal. R2, the emitter resistor, is used for proper biasing and temperature stability. C2 is a decoupling capacitor for R2. R3 develops the output signal. C3 couples the output signal to the next stage. Ql is the amplifying device.

The phantom circuit elements representing the capacitance and inductance of the wiring are: LI, L2, C4, and C5. LI represents the inductance of the input wiring. L2 represents the inductance of the output wiring. C4 represents the capacitance of the input wiring. C5 represents the capacitance of the output wiring.

In view (A) the circuit is shown with a low-frequency input signal. Since the formulas for capacitive reactance and inductive reactance are:

2n£C X L = 2n£L

You should remember that if frequency is low, capacitive reactance will be high and inductive reactance will be low. This is shown by the position of the variable resistors that represent the reactances. Notice that X L1 and X L2 are low; therefore, they do not "drop" very much of the input and output signals. X C 4 and X C 5 are high; these reactances tend to "block" the input and output signals and keep them from going to the power supplies (V B b and V C c)- Notice that the output signal is larger in amplitude than the input signal.

Now look at view (B). The input signal is a high-frequency signal. Now X c is low and X L is high. X L i and X L2 now drop part of the input and output signals. At the same time X C 4 and X C 5 tend to "short" or "pass" the input and output signals to signal ground. The net effect is that both the input and output signals are reduced. Notice that the output signal is smaller in amplitude than the input signal.

Now you can see how the capacitance and inductance of the wiring affect an amplifier, causing the output of an amplifier to be less for high-frequency signals than for low-frequency signals.

In addition to the other circuit components, an amplifying device (transistor or electronic tube), itself, reacts differently to high frequencies than it does to low frequencies. In earlier NEETS modules you were told that transistors and electronic tubes have interelectrode capacitance. Figure 2-5 shows a portion of the interelectrode capacitance of a transistor and the way in which this affects high- and low-frequency signals.

LOV FREBUENCT INPUT
'EB-T-
(A)

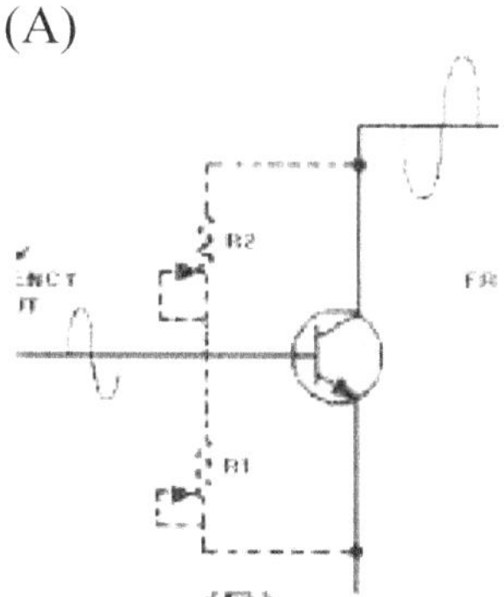

(B)
HIGH FREBUENCT INPUT
r*i.
i.—|
R1
-At
(C)

Figure 2-5.—Interelectrode capacitance of a transistor.

In view (A) a transistor is shown with phantom capacitors connected to represent the interelectrode capacitance. C E b represents the emitter-to-base capacitance. C B c represents the base-to-collector capacitance.

For simplicity, in views (B) and (C) the capacitive reactance of these capacitors is shown by variable resistors Rl (for C E b) and R2 (for C B c)- View (B) shows the reactance as high when there is a low-frequency input signal. In this case there is very little effect from the reactance on the transistor. The transistor amplifies the input signal as shown in view (B). However, when a high-frequency input signal is applied to the transistor, as in view (C), things are somewhat different. Now the capacitive reactance is low (as shown by the settings of the variable resistors). In this case, as the base of the transistor attempts to go positive during the first half of the input signal, a great deal of this positive signal is felt on the emitter (through Rl). If both the base and the emitter go positive at the same time, there is no change in emitter-base bias and the conduction of the transistor will not change. Of course, a small amount of change does occur in the emitter-base bias, but not as much as when the capacitive reactance is higher (at low frequencies). As an output signal is developed in the collector circuit, part of this signal is fed back to the base through R2. Since the signal on the collector is 180 degrees out of phase with the base signal, this tends to drive the base negative. The effect of this is to further reduce the emitter-base bias and the conduction of the transistor.During the second half of the input signal, the same effect occurs although the polarity is reversed. The net effect is a reduction in the gain of the transistor as indicated by the small output signal. This decrease in the amplifier output at higher frequencies is caused by the interelectrode capacitance. (There are certain special cases in which the feedback signal caused by the interelectrode capacitance is in phase with the base signal. However, in most cases, the feedback caused by interelectrode capacitance is degenerative and is 180 degrees out of phase with the base signal as explained above.)

Q-4. What are the factors that limit the frequency response of a transistor amplifier?

Q-5.
What type of feedback is usually caused by interelectrode capacitance?

Q-6.
What happens to capacitive reactance as frequency increases?

Q-7.
What happens to inductive reactance as frequency increases?

VIDEO AMPLIFIERS

As you have seen, a transistor amplifier is limited in its frequency response. You should also remember from chapter 1 that a VIDEO AMPLIFIER should have a frequency response of 10 hertz (10 Hz) to 6 megahertz (6 MHz). The question has probably occurred to you: How is it possible to "extend" the range of frequency response of an amplifier?

HIGH-FREQUENCY COMPENSATION FOR VIDEO AMPLIFIERS

If the frequency-response range of an audio amplifier must be extended to 6 megahertz (6 MHz) for use as a video amplifier, some means must be found to overcome the limitations of the audio amplifier. As you have seen, the capacitance of an amplifier circuit and the interelectrode capacitance of the transistor (or electronic tube) cause the higher frequency response to be limited.

In some ways capacitance and inductance can be thought of as opposites. As stated before, as frequency increases, capacitive reactance decreases, and inductive reactance increases. Capacitance opposes changes in voltage, and inductance opposes changes in current. Capacitance causes current to lead voltage, and inductance causes voltage to lead current.

Since frequency affects capacitive reactance and inductive reactance in opposite ways, and since it is the capacitive reactance that causes the problem with high-frequency response, inductors are added to an amplifier circuit to improve the high-frequency response. This is called HIGH-FREQUENCY COMPENSATION. Inductors (coils), when used for high-frequency compensation, are called PEAKING COILS. Peaking coils can be added to a circuit so they are in series with the output signal path or in parallel to the output signal path. Instead of only in series or parallel, a combination of peaking coils in series and parallel with the output signal path can also be used for high-frequency compensation.

As in all electronic circuits, nothing comes free. The use of peaking coils WILL increase the frequency response of an amplifier circuit, but it will ALSO lower the gain of the amplifier.

Series Peaking

The use of a peaking coil in series with the output signal path is known as SERIES PEAKING. Figure 2-6 shows a transistor amplifier circuit with a series peaking coil. In this figure, Rl is the input-signal-developing resistor. R2 is used for bias and temperature stability of Ql. CI is the bypass capacitor for R2. R3 is the load resistor for Ql and develops the output signal. C2 is the coupling capacitor which couples the output signal to the next stage. "Phantom" capacitor C 0 ut represents the output capacitance of the circuit, and "phantom" capacitor C IN represents the input capacitance of the next stage.

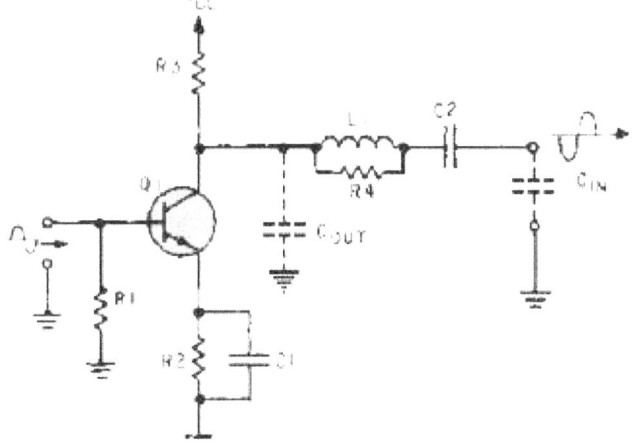

Figure 2-6.—Series peaking coil.

You know that the capacitive reactance of C 0 ut and C IN will limit the high-frequency response of the circuit. LI is the series peaking coil. It is in series with the output-signal path and isolates C 0 ut from Ci N . R4 is called a "swamping" resistor and is used to keep LI from overcompensating at a narrow range of frequencies. In other words, R4 is used to keep the frequency-response curve flat. If R4 were not used with LI, there could be a "peak" in the frequency-response curve. (Remember, LI is called a peaking coil.)

Shunt Peaking

If a coil is placed in parallel (shunt) with the output signal path, the technique is called SHUNT PEAKING. Figure 2-7 shows a circuit with a shunt peaking coil. With the exceptions of the "phantom" capacitor and the inductor, the components in this circuit are the same as those in figure 2-6. Rl is the input-signal-developing resistor. R2 is used for bias and temperature stability. CI is the bypass capacitor for R2. R3 is the load resistor for Ql and develops the output signal. C2 is the coupling capacitor which couples the output signal to the next stage.

+ vcc

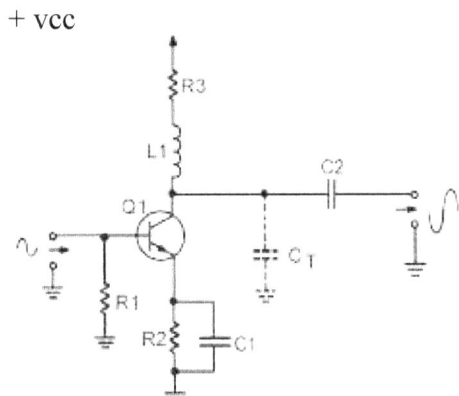

Figure 2-7.—Shunt peaking coil.

The "phantom" capacitor, C_T, represents the total capacitance of the circuit. Notice that it tends to couple the output signal to ground.

LI is the shunt peaking coil. While it is in series with the load resistor (R3), it is in parallel (shunt) with the output-signal path.

Since inductive reactance increases as frequency increases, the reactance of LI develops more output signal as the frequency increases. At the same time, the capacitive reactance of C_T is decreasing as frequency increases. This tends to couple more of the output signal to ground. The increased inductive reactance counters the effect of the decreased capacitive reactance and this increases the high-frequency response of the amplifier.

Combination Peaking

You have seen how a series peaking coil isolates the output capacitance of an amplifier from the input capacitance of the next stage. You have also seen how a shunt peaking coil will counteract the effects of the total capacitance of an amplifier. If these two techniques are used together, the combination is more effective than the use of either one alone. The use of both series and shunt peaking coils is known as COMBINATION PEAKING. An amplifier circuit with combination peaking is shown in figure 2-8. In figure 2-8 the peaking coils are LI and L2. LI is a shunt peaking coil, and L2 is a series peaking coil.

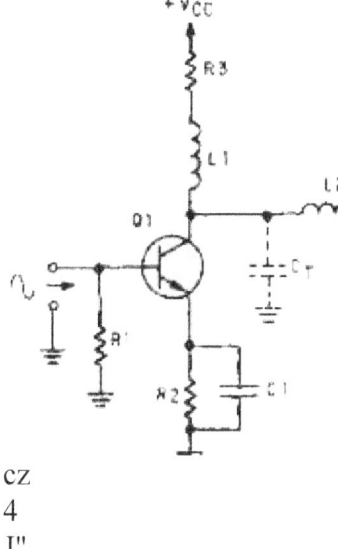

Figure 2-8.—Combination peaking.

The "phantom" capacitor C T represents the total capacitance of the amplifier circuit. "Phantom" capacitor Q N represents the input capacitance of the next stage. Combination peaking will easily allow an amplifier to have a high-frequency response of 6 megahertz (6 MHz).

Q-8. What is the major factor that limits the high-frequency response of an amplifier circuits?

Q-9. What components can be used to increase the high-frequency response of an amplifier? Q-10. What determines whether these components are considered series or shunt? Q-ll. What is the arrangement of both series and shunt components called?

LOW-FREQUENCY COMPENSATION FOR VIDEO AMPLIFIERS

Now that you have seen how the high-frequency response of an amplifier can be extended to 6 megahertz (6 MHz), you should realize that it is only necessary to extend the low-frequency response to 10 hertz (10 Hz) in order to have a video amplifier.

Once again, the culprit in low-frequency response is capacitance (or capacitive reactance). But this time the problem is the coupling capacitor between the stages.

At low frequencies the capacitive reactance of the coupling capacitor (C2 in figure 2-8) is high. This high reactance limits the amount of output signal that is coupled to the next stage. In addition, the RC network of the coupling capacitor and the signal-developing resistor of the next stage cause a phase shift in the output signal. \'7bRefer to NEETS, Module 2, for a discussion of phase shifts in RC networks.) Both of these problems (poor low-frequency response and phase shift) can be solved by adding a parallel RC network in series with the load resistor. This is shown in figure 2-9.

+ vcc

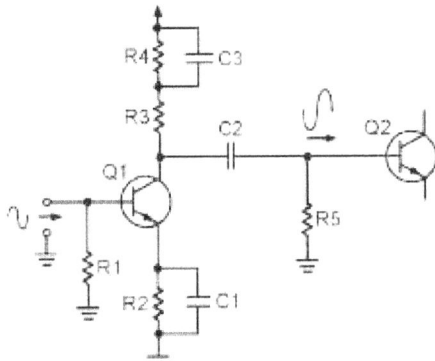

Figure 2-9.—Low frequency compensation network.

The complete circuitry for Q2 is not shown in this figure, as the main concern is the signal-developing resistor (R5) for Q2. The coupling capacitor (C2) and the resistor (R5) limit the low-frequency response of the amplifier and cause a phase shift. The amount of the phase shift will depend upon the amount of resistance and capacitance. The RC network of R4 and C3 compensates for the effects of C2 and R5 and extends the low-frequency response of the amplifier.

At low frequencies, R4 adds to the load resistance (R3) and increases the gain of the amplifier. As frequency increases, the reactance of C3 decreases. C3 then provides a path around R4 and the gain of the transistor decreases. At the same time, the reactance of the coupling capacitor (C2) decreases and more signal is coupled to Q2.

Because the circuit shown in figure 2-9 has no high-frequency compensation, it would not be a very practical video amplifier.

TYPICAL VIDEO-AMPLIFIER CIRCUIT

There are many different ways in which video amplifiers can be built. The particular configuration of a video amplifier depends upon the equipment in which the video amplifier is used. The circuit shown in figure 2-10 is only one of many possible video-amplifier circuits. Rather than reading about what each component does in this circuit, you can see how well you have learned about video amplifiers by answering the following questions. You should have no problem identifying the purpose of the components because similar circuits have been explained to you earlier in the text.

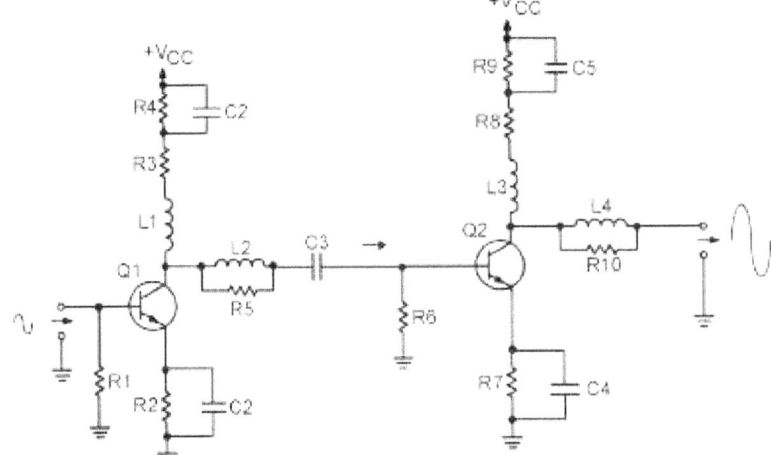

Figure 2-10.—Video amplifier circuit.

The following questions refer to figure 2-10. Q-12. What component in an amplifier circuit tends to limit the low-frequency response of the amplifier? Q-13. What is the purpose of L3? Q-14. What is the purpose of CI ? Q-15. What is the purpose of R4? Q-16. What is the purpose of L2 ? Q-17. What is the purpose of R5 ?

Q-18. What component s) is/are used for high-frequency compensation for Ql ?
Q-19. What component(s) is/are used for low-frequency compensation for Q2?

RADIO-FREQUENCY AMPLIFIERS

Now that you have seen the way in which a broadband, or video, amplifier can be constructed, you may be wondering about radio-frequency (rf) amplifiers. Do they use the same techniques? Are they just another type of broadband amplifier?

The answer to both questions is "no." Radio-frequency amplifiers use different techniques than video amplifiers and are very different from them.

Before you study the specific techniques used in rf amplifiers, you should review some information on the relationship between the input and output impedance of an amplifier and the gain of the amplifier

AMPLIFIER INPUT/OUTPUT IMPEDANCE AND GAIN

You should remember that the gain of a stage is calculated by using the input and output signals. The formula used to calculate the gain of a stage is:

Voltage gain is calculated using input and output voltage; current gain uses input and output current; and power gain uses input and output power. For the purposes of our discussion, we will only be concerned with voltage gain.

Figure 2-11 shows a simple amplifier circuit with the input- and output-signal-developing impedances represented by variable resistors. In this circuit, CI and C2 are the input and output coupling capacitors. Rl represents the impedance of the input circuit. R2 represents the input-signal-developing impedance, and R3 represents the output impedance.

Rl and R2 form a voltage-divider network for the input signal. When R2 is increased in value, the input signal to the transistor (Ql) increases. This causes a larger output signal, and the gain of the stage increases.

Now look at the output resistor, R3. As R3 is increased in value, the output signal increases. This also increases the gain of the stage.

As you can see, increasing the input-signal-developing impedance, the output impedance, or both will increase the gain of the stage. Of course there are limits to this process. The transistor must not be overdriven with too high an input signal or distortion will result.

With this principle in mind, if you could design a circuit that had maximum impedance at a specific frequency (or band of frequencies), that circuit could be used in an rf amplifier. This FREQUENCY-DETERMINING NETWORK could be used as the input-signal-developing impedance, the output impedance, or both. The rf amplifier circuit would then be as shown in figure 2-12.

stage.
gam =
Output Signal Input Signal

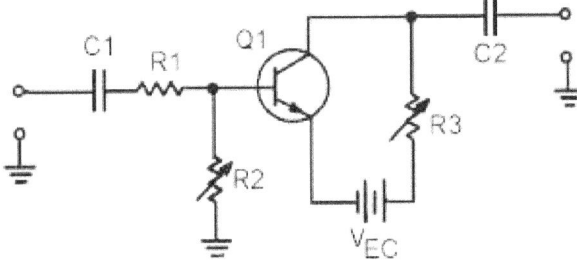

Figure 2-11.—Variable input and output impedances.

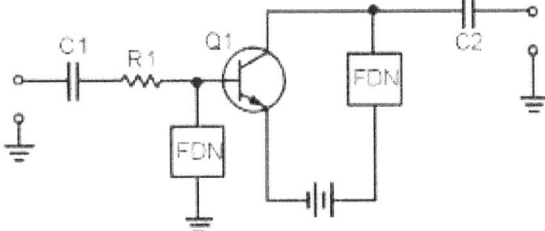

Figure 2-12.—Semiblock diagram of rf amplifier.

In this "semi-block" diagram, CI and C2 are the input and output coupling capacitors. Rl represents the impedance of the input circuit. The blocks marked FDN represent the frequency-determining networks. They are used as input-signal-developing and output impedances for Ql.

FREQUENCY-DETERMINING NETWORK FOR AN RF AMPLIFIER

What kind of circuit would act as a frequency-determining network? In general, a frequency-determining network is a circuit that provides the desired response at a particular frequency. This response could be maximum impedance or minimum impedance; it all depends on how the frequency-determining network is used. You will see more about frequency-determining networks in NEETS, Module 9 — Introduction to Wave-Generation and Shaping Circuits. As you have seen, the frequency-determining network needed for an rf amplifier should have maximum impedance at the desired frequency.

Before you are shown the actual components that make up the frequency-determining network for an rf amplifier, look at figure 2-13, which is a simple parallel circuit. The resistors in this circuit are variable and are connected together (ganged) in such a way that as the resistance of Rl increases, the resistance of R2 decreases, and vice versa.

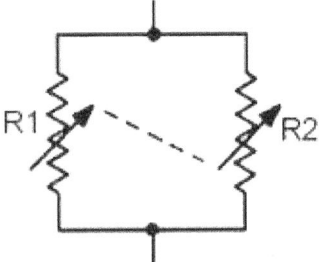

Figure 2-13.—Parallel variable resistors (ganged).

If each resistor has a range from 0 to 200 ohms, the following relationship will exist between the individual resistances and the resistance of the network (R T)- (All values are in ohms, R T rounded off to two decimal places. These are selected values; there are an infinite number of possible combinations.)

As you can see, this circuit has maximum resistance (R T) when the individual resistors are of equal value. If the variable resistors represented impedances and if components could be found that varied their impedance in the same way as the ganged resistors in figure 2-13, you would have the frequency-determining network needed for an rf amplifier.

There are components that will vary their impedance (reactance) like the ganged resistors. As you know, the reactance of an inductor and a capacitor vary as frequency changes. As frequency increases, inductive reactance increases, and capacitive reactance decreases.

At some frequency, inductive and capacitive reactance will be equal. That frequency will depend upon the value of the inductor and capacitor. If the inductor and capacitor are connected as a parallel LC circuit, you will have the ideal frequency-determining network for an rf amplifier.

The parallel LC circuit used as a frequency-determining network is called a TUNED CIRCUIT. This circuit is "tuned" to give the proper response at the desired frequency by selecting the proper values of inductance and capacitance. A circuit using this principle is shown in figure 2-14 which shows an rf amplifier with parallel LC circuits used as frequency-determining networks. This rf amplifier will only be effective in amplifying the frequency determined by the parallel LC circuits.

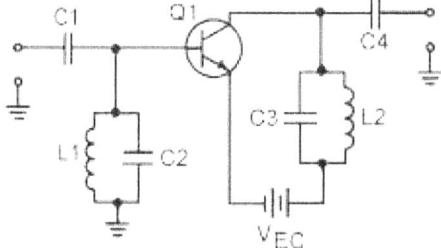

Figure 2-14.—Simple rf amplifier.

In many electronic devices, such as radio or television receivers or radar systems, a particular frequency must be selected from a band of frequencies. This could be done by using a separate rf amplifier for each frequency and then turning on the appropriate rf amplifier. It would be more efficient if a single rf amplifier could be "tuned" to the particular frequency as that frequency is needed. This is what

happens when you select a channel on your television set or tune to a station on your radio. To accomplish this "tuning," you need only change the value of inductance or capacitance in the parallel LC circuits (tuned circuits).

In most cases, the capacitance is changed by the use of variable capacitors. The capacitors in the input and output portions of all the rf amplifier stages are ganged together in order that they can all be changed at one time with a single device, such as the tuning dial on a radio. (This technique will be shown on a schematic a little later in this chapter.)

Q-20. If the input-signal-developing impedance of an amplifier is increased, what is the effect on the gain?

Q-21. If the output impedance of an amplifier circuit is decreased, what is the effect on the gain?

Q-22. What is the purpose of a frequency-determining network in an rf amplifier?

Q-23. Can a parallel LC circuit be used as the frequency-determining network for an rf amplifier?

Q-24. How can the frequency be changed in the frequency-determining network?

RF AMPLIFIER COUPLING

Figure 2-14 and the other circuits you have been shown use capacitors to couple the signal in to and out of the circuit (CI and C4 in figure 2-14). As you remember from chapter 1, there are also other methods of coupling signals from one stage to another. Transformer coupling is the most common method used to couple rf amplifiers. Transformer coupling has many advantages over RC coupling for rf amplifiers; for example, transformer coupling uses fewer components than capacitive coupling. It can also provide a means of increasing the gain of the stage by using a step-up transformer for voltage gain. If a current gain is required, a step-down transformer can be used.

You should also remember that the primary and secondary windings of a transformer are inductors. With these factors in mind, an rf amplifier could be constructed like the one shown in figure 2-15.

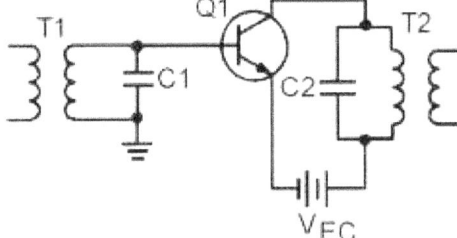

Figure 2-15.—Transformer-coupled rf amplifier.

In this circuit, the secondary of T1 and capacitor CI form a tuned circuit which is the input-signal-developing impedance. The primary of T2 and capacitor C2 are a tuned circuit which acts as the output impedance of Ql. (Both Tl and T2 must be rf transformers in order to operate at rf frequencies.)

The input signal applied to the primary of T1 could come from the previous stage or from some input device, such as a receiving antenna. In either case, the input device would have a capacitor connected

across a coil to form a tuned circuit. In the same way, the secondary of T2 represents the output of this circuit. A capacitor connected across the secondary of T2 would form a parallel LC network. This network could act as the input-signal-developing impedance for the next stage, or the network could represent some type of output device, such as a transmitting antenna.

The tuned circuits formed by the transformer and capacitors may not have the bandwidth required for the amplifier. In other words, the bandwidth of the tuned circuit may be too "narrow" for the requirements of the amplifier. (For example, the rf amplifiers used in television receivers usually require a bandwidth of 6 MHz.)

One way of "broadening" the bandpass of a tuned circuit is to use a swamping resistor. This is similar to the use of the swamping resistor that was shown with the series peaking coil in a video amplifier. A swamping resistor connected in parallel with the tuned circuit will cause a much broader bandpass. (This technique and the theory behind it are discussed in more detail in NEETS, Module 9.)

Another technique used to broaden the bandpass involves the amount of coupling in the transformers. For transformers, the term "coupling" refers to the amount of energy transferred from the primary to the secondary of the transformer. This depends upon the number of flux lines from the primary that intersect, or cut, the secondary. When more flux lines cut the secondary, more energy is transferred.

Coupling is mainly a function of the space between the primary and secondary windings. A transformer can be loosely coupled (having little transfer of energy), optimumly coupled (just the right amount of energy transferred), or overcoupled (to the point that the flux lines of primary and secondary windings interfere with each other).

Figure 2-16, (view A) (view B) (view C), shows the effect of coupling on frequency response when parallel LC circuits are made from the primary and secondary windings of transformers.

A. LOOSE COUPLING FREQUENCY

Figure 2-16A.—Effect of coupling on frequency response. LOOSE COUPLING

B. OPTIMUM COUPLING FREQUENCY

Figure 2-16B.—Effect of coupling on frequency response. OPTIMUM COUPLING

C. OVER-COUPLING FREQUENCY

Figure 2-16C—Effect of coupling on frequency response. OVER-COUPLING

In view (A) the transformer is loosely coupled; the frequency response curve shows a narrow bandwidth. In view (B) the transformer has optimum coupling; the bandwidth is wider and the curve is relatively flat. In view (C) the transformer is overcoupled; the frequency response curve shows a broad bandpass, but the curve "dips" in the middle showing that these frequencies are not developed as well as others in the bandwidth.

Optimum coupling will usually provide the necessary bandpass for the frequency-determining network (and therefore the rf amplifier). For some uses, such as rf amplifiers in a television receiver, the bandpass available from optimum coupling is not wide enough. In these cases, a swamping resistor (as mentioned earlier) will be used with the optimum coupling to broaden the bandpass.

COMPENSATION OF RF AMPLIFIERS

Now you have been shown the way in which an rf amplifier is configured to amplify a band of frequencies and the way in which an rf amplifier can be "tuned" for a particular band of frequencies. You have also seen some ways in which the bandpass of an rf amplifier can be adjusted. However, the frequencies at which rf amplifiers operate are so high that certain problems exist.

One of these problems is the losses that can occur in a transformer at these high frequencies. Another problem is with interelectrode capacitance in the transistor. The process of overcoming these problems is known as COMPENSATION.

Transformers in RF Amplifiers

As you recall from NEETS, Module 1, me losses in a transformer are classified as copper loss, eddy-current loss, and hysteresis loss. Copper loss is not affected by frequency, as it depends upon the resistance of the winding and the current through the winding. Similarly, eddy-current loss is mostly a function of induced voltage rather than the frequency of that voltage. Hysteresis loss, however, increases as frequency increases.

Hysteresis loss is caused by the realignment of the magnetic domains in the core of the transformer each time the polarity of the magnetic field changes. As the frequency of the a.c. increases, the number of shifts in the magnetic field also increases (two shifts for each cycle of a.c); therefore, the "molecular friction" increases and the hysteresis loss is greater. This increase in hysteresis loss causes the efficiency of the transformer (and therefore the amplifier) to decrease. The energy that goes into hysteresis loss is taken away from energy that could go into the signal.

RF TRANSFORMERS, specially designed for use with rf, are used to correct the problem of excessive hysteresis loss in the transformer of an rf amplifier. The windings of rf transformers are wound onto a tube of nonmagnetic material and the core is either powdered iron or air. These types of cores also reduce eddy-current loss.

Neutralization of RF Amplifiers

The problem of interelectrode capacitance in the transistor of an rf amplifier is solved by NEUTRALIZATION. Neutralization is the process of counteracting or "neutralizing" the effects of interelectrode capacitance.

Figure 2-17 shows the effect of the base-to-collector interelectrode capacitance in an rf amplifier. The "phantom" capacitor (C B c) represents the interelectrode capacitance between the base and the collector of Ql. This is the interelectrode capacitance that has the most effect in an rf amplifier. As you can see, C B c causes a degenerative (negative) feedback which decreases the gain of the amplifier. (There are some special cases in which C B c can cause regenerative (positive) feedback. In this case, the technique described below will provide negative feedback which will accomplish the neutralization of the amplifier.)

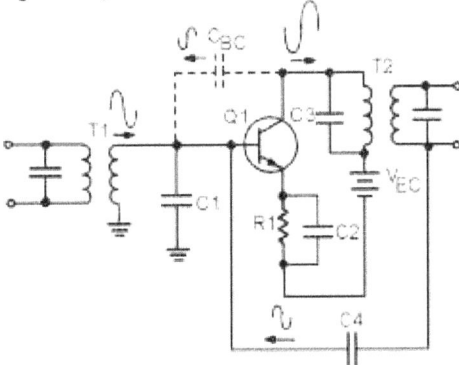

Figure 2-17.—Interelectrode capacitance in an rf amplifier.

As you may recall, unwanted degenerative feedback can be counteracted (neutralized) by using positive feedback. This is exactly what is done to neutralize an rf amplifier.

Positive feedback is accomplished by the use of a feedback capacitor. This capacitor must feed back a signal that is in phase with the signal on the base of Ql. One method of doing this is shown in figure 2-18.

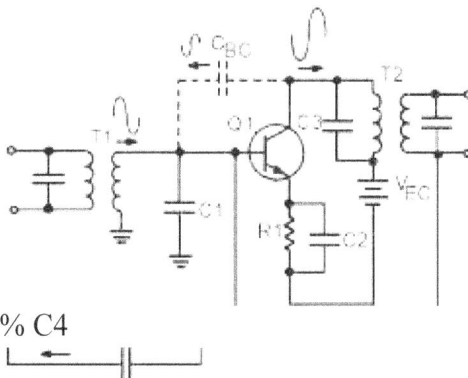

Figure 2-18.—Neutralized rf amplifier.

In figure 2-18, a feedback capacitor (C4) has been added to neutralize the amplifier. This solves the problem of unwanted degenerative feedback. Except for capacitor C4, this circuit is identical to the circuit shown in figure 2-17. (When C B c causes regenerative feedback, C4 will still neutralize the amplifier. This is true because C4 always provides a feedback signal which is 180 degrees out of phase with the feedback signal caused by C B c-)

Q-25. What is the most common form of coupling for an rf amplifier? Q-26. What are two advantages of this type of coupling?

Q-27. If current gain is required from an rf amplifier, what type of component should be used as an output coupling element?

Q-28. What problem is caused in an rf amplifier by a loosely coupled transformer?

Q-29. How is this problem corrected?

Q-30. What problem is caused by overcoupling in a transformer?

Q-31. What method provides the widest bandpass?

Q-32. What two methods are used to compensate for the problems that cause low gain in an rf amplifier?

Q-33. What type of feedback is usually caused by the base-to-collector interelectrode capacitance?

Q-34. How is this compensated for?

TYPICAL RF AMPLIFIER CIRCUITS

As a technician, you will see many different rf amplifiers in many different pieces of equipment. The particular circuit configuration used for an rf amplifier will depend upon how that amplifier is used. In the final part of this chapter, you will be shown some typical rf amplifier circuits.

Figure 2-19 is the schematic diagram of a typical rf amplifier that is used in an AM radio receiver. In figure 2-19, the input circuit is the antenna of the radio (Ll-a coil) which forms part of an LC circuit which is tuned to the desired station by variable capacitor CI. LI is wound on the same core as L2, which couples the input signal through C2 to the transistor (Ql). Rl is used to provide proper bias to Ql from the base power supply (V B b)- R2 provides proper bias to the emitter of Ql, and C3 is used to bypass R2. The primary of Tl and capacitor C4 form a parallel LC circuit which acts as the load for Ql. This LC circuit is tuned by C4, which is ganged to CI allowing the antenna and the LC circuit to be tuned together. The primary of Tl is center-tapped to provide proper impedance matching with Ql.

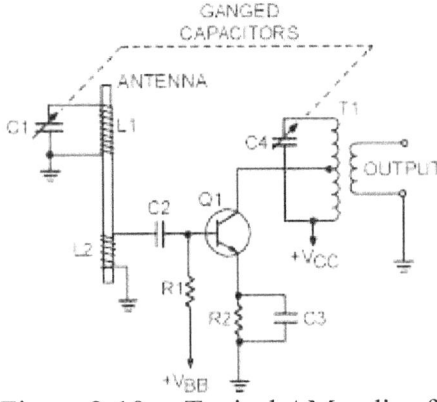

Figure 2-19.—Typical AM radio rf amplifier.

You may notice that no neutralization is shown in this circuit. This circuit is designed for the AM broadcast band (535 kHz - 1605 kHz).

At these relatively low rf frequencies the degenerative feedback caused by base-to-collector interelectrode capacitance is minor and, therefore, the amplifier does not need neutralization.

Figure 2-20 is a typical rf amplifier used in a vhf television receiver. The input-signal-developing circuit for this amplifier is made up of LI, CI, and C2. The inductor tunes the input-signal-developing circuit for the proper TV channel. (LI can be switched out of the circuit and another inductor switched in to the circuit by the channel selector.) Rl provides proper bias to Ql from the base supply voltage (V B b)-Ql is the transistor. Notice that the case of Ql (the dotted circle around the transistor symbol) is shown to be grounded. The case must be grounded because of the high frequencies (54 MHz - 217 MHz) used by the circuit. R2 provides proper bias from the emitter of Ql, and C3 is used to bypass R2. C5 and L2 are a parallel LC circuit which acts as the load for Ql. The LC circuit is tuned by L2 which is switched in to and out of the LC circuit by the channel selector. L3 and C6 are a parallel LC circuit which develops the signal for the next stage. The parallel LC circuit is tuned by L3 which is switched in to and out of the LC circuit by the channel selector along with LI and L2. (LI, L2, and L3 are actually part of a bank of inductors. LI, L2, and L3 are in the circuit when the channel selector is on channel 2. For other channels, another group of three inductors would be used in the circuit.) R3 develops a signal which is fed through C4 to provide neutralization. This counteracts the effects of the interelectrode capacitance from the base to the collector of Ql. C7 is used to isolate the rf signal from the collector power supply (V C c)-

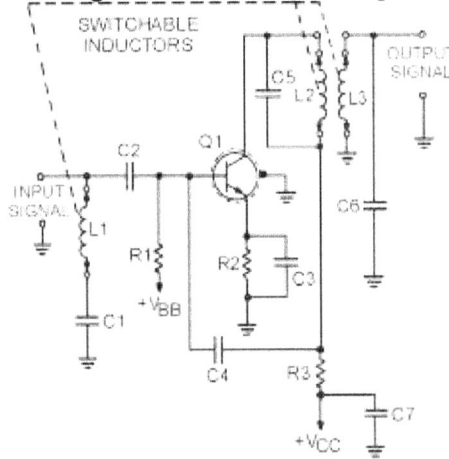

Figure 2-20.—Typical vhf television rf amplifier.

The following questions refer to figure 2-21.

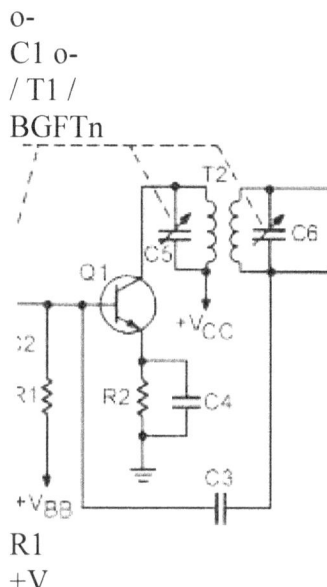

Figure 2-21.—Typical rf amplifier.

Q-35. What components form the input-signal-developing impedance for the amplifier? Q-36. What is the purpose of Rl?

Q-37. What is the purpose of R2?

Q-38. If C4 were removed from the circuit, what would happen to the output of the amplifier?

Q-39. What components form the load for Ql?

Q-40. How many tuned parallel LC circuits are shown in this schematic?

Q-41. What do the dotted lines connecting CI, C2, C5, and C6 indicate?

Q-42. What is the purpose of C3?

SUMMARY

This chapter has presented information on video and rf amplifiers. The information that follows summarizes the important points of this chapter.

A FREQUENCY-RESPONSE CURVE will enable you to determine the BANDWIDTH and the UPPER and LOWER FREQUENCY LIMITS of an amplifier.

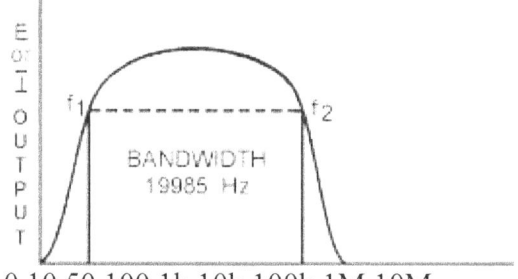

0 10 50 100 1k 10k 100k 1M 10M
FREQUENCY (Hz)

The BANDWIDTH of an amplifier is determined by the formula:

BW = f2 - fi

Where:

BW is the bandwidth fj is the upper-frequency limit and

fl is the lower-frequency limit

The UPPER-FREQUENCY RESPONSE of an amplifier is limited by the inductance and capacitance of the circuit.

HIGH n ^ p ._tJ,,, FREQUENCY | ; i ai
LI
CI
R1
cc
: i
BB
i
.1.

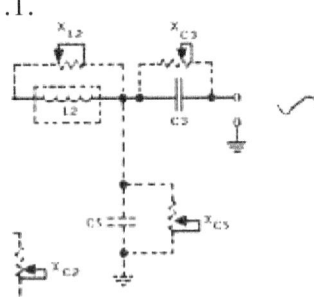

The INTERELECTRODE CAPACITANCE of a transistor causes DEGENERATIVE FEEDBACK at high frequencies.

T~t-\
HIGH J "V* R2
FREQUENCY « S INPUT - I J
•i

VIDEO AMPLIFIERS must have a frequency response of 10 hertz to 6 megahertz (10 Hz - 6 MHz). To provide this frequency response, both high- and low-frequency compensation must be used.

PEAKING COILS are used in video amplifiers to overcome the high-frequency limitations caused by the capacitance of the circuit.

SERIES PEAKING is accomplished by a peaking coil in series with the output-signal path.

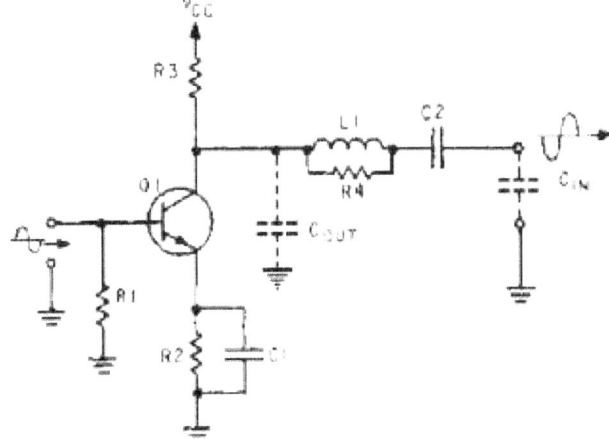

SHUNT PEAKING is accomplished by a peaking coil in parallel (shunt) with the output-signal path.

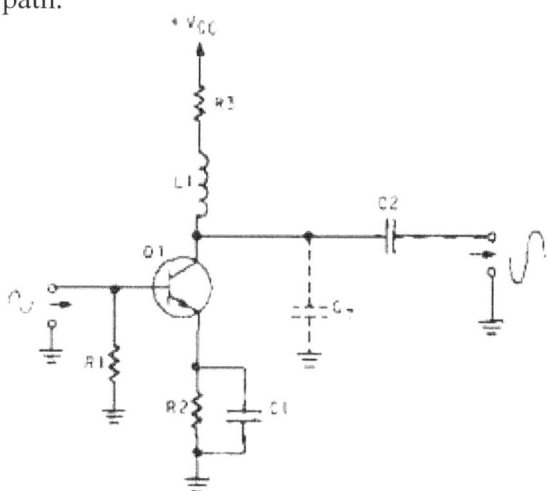

COMBINATION PEAKING is accomplished by using both series and shunt peaking.

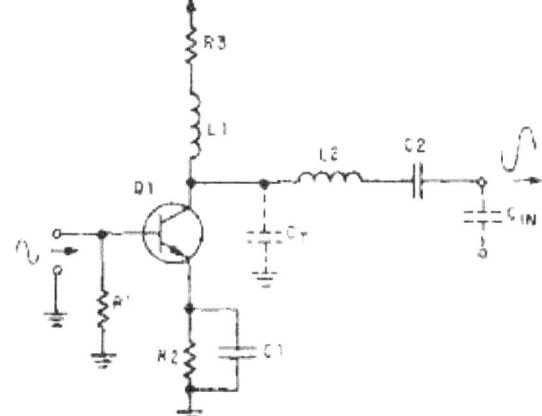

LOW-FREQUENCY COMPENSATION is accomplished in a video amplifier by the use of a parallel RC circuit in series with the load resistor.

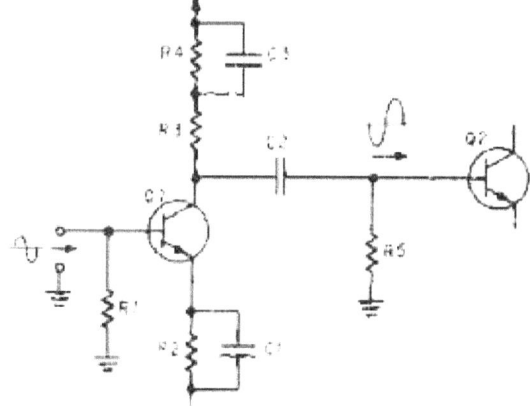

A RADIO-FREQUENCY (RF) AMPLIFIER uses FREQUENCY-DETERMINING NETWORKS to provide the required response at a given frequency.

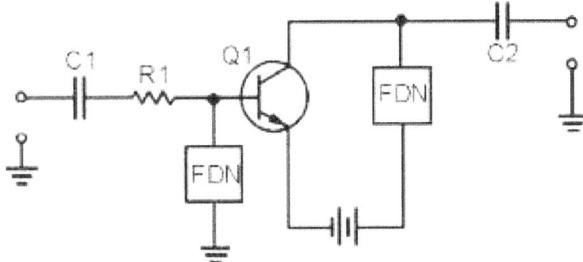

The FREQUENCY-DETERMINING NETWORK in an rf amplifier provides maximum impedance at the desired frequency. It is a parallel LC circuit which is called a TUNED CIRCUIT.

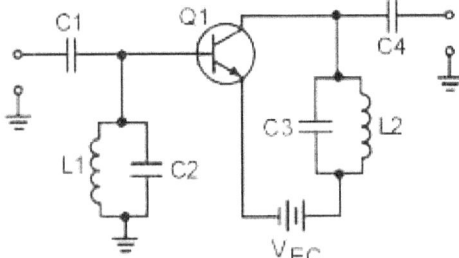

TRANSFORMER COUPLING is the most common form of coupling in rf amplifiers. This coupling is accomplished by the use of rf transformers as part of the frequency-determining network for the amplifier.

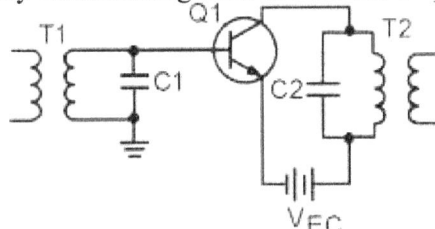

ADEQUATE BANDPASS is accomplished by optimum coupling in the rf transformer or by the use of a SWAMPING RESISTOR.

NEUTRALIZATION in an rf amplifier provides feedback (usually positive) to overcome the effects caused by the base-to-collector interelectrode capacitance.

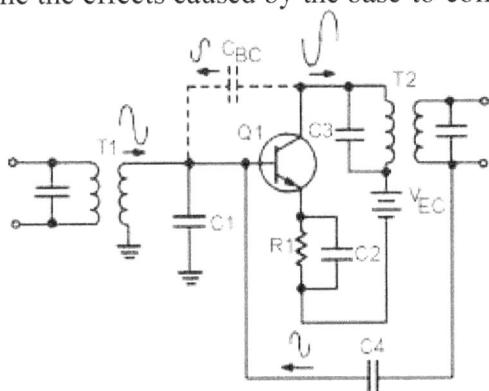

ANSWERS TO QUESTIONS Q1. THROUGH Q42.

A-1. The difference between the upper and lower frequency limits of an amplifier.

A-2. The half-power points of a frequency-response curve. The upper and lower limits of the bandf frequencies for which the amplifier is most effective.

A-3. (A)$f_2 = 80$ kHz, $f_i = 30$ kHz, BW = 50 kHz (B)$f_2 = 4$ kHz, $f_i = 2$ kHz,

BW=2 kHz

A-4. The capacitance and inductance of the circuit and the interelectrode capacitance of the transistor.

A-5. Negative (degenerative) feedback.

A-6. It decreases.

A-7. It increases.

A-8. The capacitance of the circuit.

A-9. Peaking coils.

A-10. The relationship of the components to the output-signal path.

A-ll. Combination peaking.

A-12. The coupling capacitor (C3).

A-13. A shunt peaking coil for Q2.

A-14. A decoupling capacitor for the effects ofR2.

A-15. A part of the low-frequency compensation network for Ql.

A-16. A series peaking coil for Ql.

A-l 7. A swamping resistor for 12.

A-18. Ll,L2,andR5.

A-19. R9 and C5.

A-20. The gain increases.

A-21. The gain decreases.

A-22. To provide maximum impedance at the desired frequency.

A-23. Yes.

A-24. By changing the value.

A-25. Transformer coupling.

A-26. It uses fewer components than capacitive coupling and can provide an increase in gain.

A-27. A step-down transformer.

A-28. A too-narrow bandpass.

A-29. By using an optimumly-coupled transformer.

A-30. Low gain at the center frequency.

A-31. A swamping resistor in parallel with the tuned circuit.

A-32. RF transformers are used and the transistor is neutralized.

A-33. Degenerative or negative.

A-34. By neutralization such as the use of a capacitor to provide regenerative (positive) feedback.

A-35. C2 and the secondary of Tl.

A-36. Rl provides the proper bias to the base of Ql from V B b-

A-37. R2 provides the proper bias to the emitter of Ql.

A-38. The output would decrease. (C4 decouples R2 preventing degenerative feedback from R2.)

A-39. C5 and the primary of T2.

A-40. Four.

A-41. The dotted lines indicate that these capacitors are "ganged" and are tuned together with a single control.

A-42. C3 provides neutralization for Ql.

CHAPTER 3

SPECIAL AMPLIFIERS

LEARNING OBJECTIVES

Upon completion of this chapter, you will be able to:

1. Describe the basic operation of a differential amplifier.
2. Describe the operation of a differential amplifier under the following conditions:
   a. Single Input, Single Output
   b. Single input, differential output
   c. Differential input, differential output
3. List the characteristics of an operational amplifier.
4. Identify the symbol for an operational amplifier.
5. Label the blocks on a block diagram of an operational amplifier.
6. Describe the operation of an operational amplifier with inverting and noninverting configurations.
7. Describe the bandwidth of a typical operational amplifier and methods to modify the bandwidth.
8. Identify the following applications of operational amplifiers:
   a. Adder
   b. Subtractor
9. State the common usage for a magnetic amplifier.
10. Describe the basic operation of a magnetic amplifier.
11. Describe various methods of changing inductance.
12. Identify the purpose of components in a simple magnetic amplifier.

INTRODUCTION

If you were to make a quick review of the subjects discussed in this module up to this point, you would see that you have been given a considerable amount of information about amplifiers. You have been shown what amplification is and how the different classes of amplifiers affect amplification. You also have been shown that many factors must be considered when working with amplifiers, such as impedance, feedback, frequency response, and coupling. With all this information behind you, you might ask yourself "what more can there be to know about amplifiers?"

There is a great deal more to learn about amplifiers. Even after you finish this chapter you will have only "scratched the surface" of the study of amplifiers. But, you will have prepared yourself for the remainder of the NEETS. This, in turn, should prepare you for further study and, perhaps, a career in electronics.

As in chapter 2, the circuits shown in this chapter are intended to present particular concepts to you. Therefore, the circuits may be incomplete or not practical for use in an actual piece of electronic equipment. You should keep in mind the fact that this text is intended to teach certain facts about amplifiers, and in order to simplify the illustrations used, complete operational circuits are not always shown.

In this chapter three types of special amplifiers are discussed. These are: DIFFERENTIAL AMPLIFIERS, OPERATIONAL AMPLIFIERS, and MAGNETIC AMPLIFIERS. These are called special amplifiers because they are used only in certain types of equipment.

The names of each of these special amplifiers describe the operation of the amplifier, NOT what is amplified. For example, a magnetic amplifier does not amplify magnetism but uses magnetic effects to produce amplification of an electronic signal.

A differential amplifier is an amplifier that can have two input signals and/or two output signals. This amplifier can amplify the difference between two input signals. A differential amplifier will also "cancel out" common signals at the two inputs.

One of the more interesting aspects of an operational amplifier is that it can be used to perform mathematical operations electronically. Properly connected, an operational amplifier can add, subtract, multiply, divide, and even perform the calculus operations of integration and differentiation. These amplifiers were originally used in a type of computer known as the "analog computer" but are now used in many electronic applications.

The magnetic amplifier uses a device called a "saturable core reactor" to control an a.c.output signal. The primary use of magnetic amplifiers is in power control systems.

These brief descriptions of the three special amplifiers are intended to provide you with a general idea of what these amplifiers are and how they can be used. The remaining sections of this chapter will provide you with more detailed information on these special amplifiers.

DIFFERENTIAL AMPLIFIERS

A differential amplifier has two possible inputs and two possible outputs. This arrangement means that the differential amplifier can be used in a variety of ways. Before examining the three basic configurations that are possible with a differential amplifier, you need to be familiar with the basic circuitry of a differential amplifier.

BASIC DIFFERENTIAL AMPLIFIER CIRCUIT

Before you are shown the operation of a differential amplifier, you will be shown how a simpler circuit works. This simpler circuit, known as the DIFFERENCE AMPLIFIER, has one thing in common

with the differential amplifier: It operates on the difference between two inputs . However, the difference amplifier has only one output while the differential amplifier can have two outputs.

By now, you should be familiar with some amplifier circuits, which should give you an idea of what a difference amplifier is like. In NEETS, Module 7, you were shown the basic configurations for transistor amplifiers. Figure 3-1 shows two of these configurations: the common emitter and the common base.

In view (A) of figure 3-1 a common-emitter amplifier is shown. The output signal is an amplified version of the input signal and is 180 degrees out of phase with the input signal. View (B) is a common-base amplifier. In this circuit the output signal is an amplified version of the input signal and is in phase with the input signal. In both of these circuits, the output signal is controlled by the base-to-emitter bias. As this bias changes (because of the input signal) the current through the transistor changes. This causes the output signal developed across the collector load (R2) to change. None of this information is new, it is just a review of what you have already been shown regarding transistor amplifiers.

+V CC

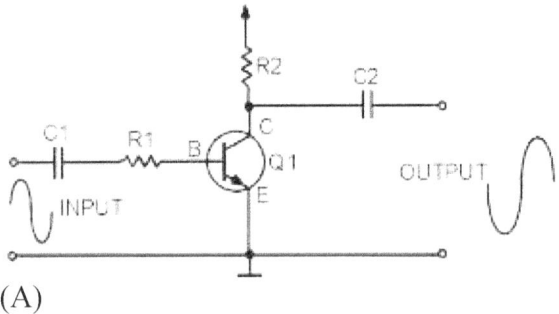

(A)

Figure 3-1A.—Common-emitter and common-base amplifiers.

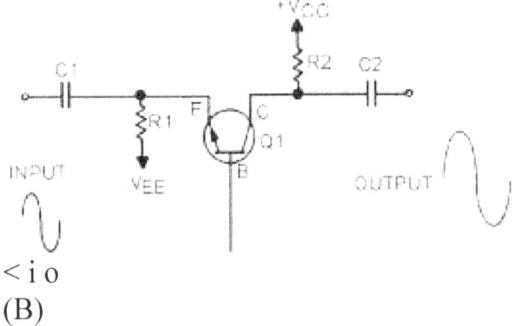

(B)

Figure 3-1B.—Common-emitter and common-base amplifiers.

NOTE: Bias arrangements for the following explanations will be termed base-to-emitter. In other publications you will see the term emitter-to-base used to describe the same bias arrangement.

THE TWO-INPUT, SINGLE-OUTPUT, DIFFERENCE AMPLIFIER

If you combine the common-base and common-emitter configurations into a single transistor amplifier, you will have a circuit like the one shown in figure 3-2. This circuit is the two-input, single-output, difference amplifier.

In figure 3-2, the transistor has two inputs (the emitter and the base) and one output (the collector). Remember, the current through the transistor (and therefore the output signal) is controlled by the base-to-emitter bias. In the circuit shown in figure 3-2, the combination of the two input signals controls the output signal. In fact, the DIFFERENCE BETWEEN THE INPUT SIGNALS determines the base-to-emitter bias.

For the purpose of examining the operation of the circuit shown in figure 3-2, assume that the circuit has a gain of -10. This means that for each 1-volt change in the base-to-emitter bias, there would be a 10-volt change in the output signal. Assume, also, that the input signals will peak at 1-volt levels (+1 volt for the positive peak and -1 volt for the negative peak). The secret to understanding this circuit (or any transistor amplifier circuit) is to realize that the collector current is controlled by the base-to-emitter bias. In other words, in this circuit the output signal (the voltage developed across R3) is determined by the difference between the voltage on the base and the voltage on the emitter.

Figure 3-3 shows this two-input, single-output amplifier with input signals that are equal in amplitude and 180 degrees out of phase. Input number one has a positive alternation when input number two has a negative alternation and vice versa.

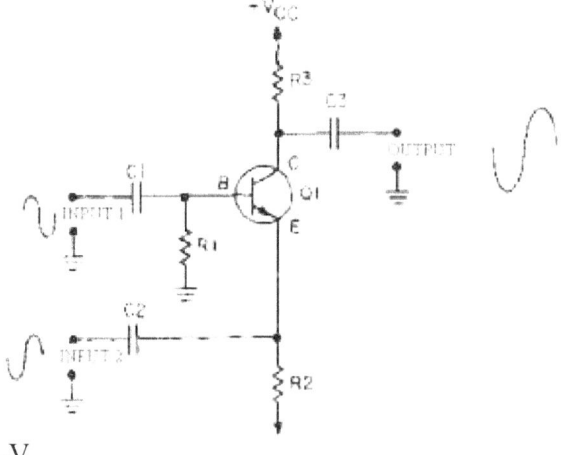

Figure 3-2.—Two-input, single-output, difference amplifier.

INFUT*1 "T 1
—*H

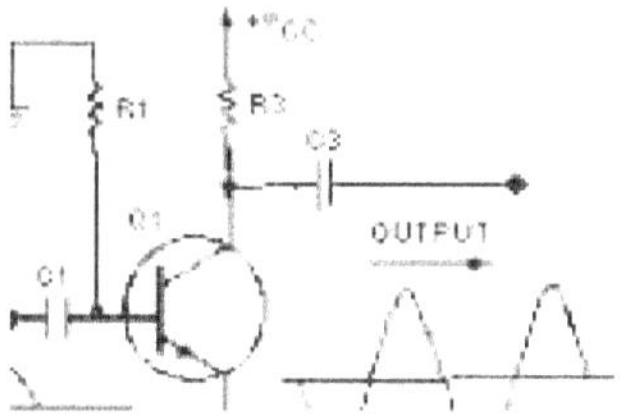

INPUT *1
C2
v v i
RE
TO T1 T2 T3 T4 T5 Tf- TT Tt
+ 1Y
INFUTtl 0
- 1Y
+ 1Y
INPUT *2 0
- 1Y
+ 2Y + 1Y OUTPUT 0
- 1Y
- £Y

Figure 3-3.—Input signals 180° out of phase.

The circuit and the input and output signals are shown at the top of the figure. The lower portion of the figure is a comparison of the input signals and the output signal. Notice the vertical lines marked "TO" through "T8." These represent "time zero" through "time eight." In other words, these lines provide a way to examine the two input signals and the output signal at various instants of time.

3-5

In figure 3-3 at time zero (TO) both input signals are at 0 volts. The output signal is also at 0 volts. Between time zero (TO) and time one (Tl), input signal number one goes positive and input signal number two goes negative. Each of these voltage changes causes an increase in the base-to-emitter bias which causes current through Ql to increase. Increased current through Ql results in a greater voltage drop across the collector load (R3) which causes the output signal to go negative.

By time one (Tl), input signal number one has reached +1 volt and input signal number two has reached -1 volt. This is an overall increase in base-to-emitter bias of 2 volts. Since the gain of the circuit is -10, the output signal has decreased by 20 volts. As you can see, the output signal has been determined by the difference between the two input signals. In fact, the base-to-emitter bias can be found by subtracting the value of input signal number two from the value of input signal number one.

Mathematically:
Bias = (input signal #1) - (input signal #2) Bias = (+lV)-(-lV) Bias = +1V+ IV Bias = +2V

Between time one (Tl) and time two (T2), input signal number one goes from +1 volt to 0 volts and input signal number two goes from -1 volt to 0 volts. At time two (T2) both input signals are at 0 volts and the base-to-emitter bias has returned to 0 volts. The output signal is also 0 volts.

Mathematically:
Bias = (input signal #1) - (input signal #2) Bias = (OV) - (OV) Bias = OV

Between time two (T2) and time three (T3), input signal number one goes negative and input signal number two goes positive. At time three (T3), the value of the base-to-emitter bias is -2 volts.

Mathematically:
Bias = (input signal #1) - (input signal #2) Bias = (-1V)- (+1V) Bias = (-1V) + (-1V) Bias = -2V

This causes the output signal to be +20 volts at time three (T3).

Between time three (T3) and time four (T4), input signal #1 goes from -1 volt to 0 volts and input signal #2 goes from +1 volt to 0 volts. At time four (T4) both input signals are 0 volts, the bias is 0 volts, and the output is 0 volts.

During time four (T4) through time eight (T8), the circuit repeats the sequence of events that took place from time zero (TO) through time four (T4).

You can see that when the input signals are equal in amplitude and 180 degrees out of phase, the output signal is twice as large (40 volts peak to peak) as it would be from either input signal alone (if the other input signal were held at 0 volts).

Figure 3-4 shows the two-input, single-output, difference amplifier with two input signals that are equal in amplitude and in phase.

C3
OUTPUT *
INPUT *1 ci /
JiRl
INPUTtE CE
TO T1 TE TS T4 T5 Tt TT T*
INPUT i

Figure 3-4.—Input signals in phase.

Notice, that the output signal remains at 0 volts for the entire time (TO - T8). Since the two input signals are equal in amplitude and in phase, the difference between them (the base-to-emitter bias) is always 0 volts. This causes a 0-volt output signal.

If you compute the bias at any time period (TO - T8), you will see that the output of the circuit remains at a constant zero.

For example:

Bias = (input signal #1) - (input signal #2) Tl Bias = (+1V)- (+1V) = 0 and so forth

From the above example, you can see that when the input signals are equal in amplitude and in phase, there is no output from the difference amplifier because there is no difference between the two inputs. You also know that when the input signals are equal in amplitude but 180 degrees out of phase, the output looks just like the input except for amplitude and a 180-degree phase reversal with respect to input signal number one. What happens if the input signals are equal in amplitude but different in phase by something other than 180 degrees? This would mean that sometimes one signal would be going negative while the other would be going positive; sometimes both signals would be going positive; and sometimes both signals would be going negative. Would the output signal still look like the input signals? The answer is "no," because figure 3-5 shows a difference amplifier with two input signals that are equal in amplitude but 90 degrees out of phase. From the figure you can see that at time zero (TO) input number one is at 0 volts and input number two is at -1 volt. The base-to-emitter bias is found to be +1 volt.

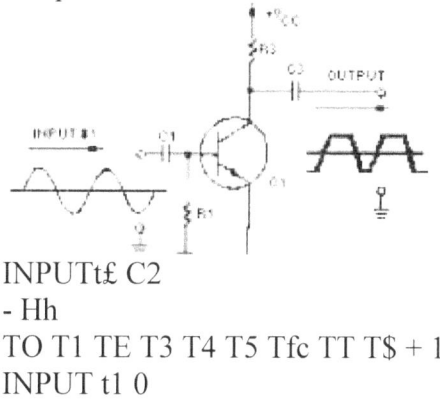

INPUTt£ C2
- Hh
TO T1 TE T3 T4 T5 Tfc TT T$ + 1Y
INPUT tl 0
- 1Y
+ W
INPUT t£ 0
- 1Y
OUTPUT <; - 10V-

Figure 3-5.—Input signals 90° out of phase.

This +l-volt bias signal causes the output signal to be -10 volts at time zero (TO). Between time zero (TO) and time one (Tl), both input signals go positive. The difference between the input signals stays constant. The effect of this is to keep the bias at +1 volt for the entire time between TO and Tl. This, in turn, keeps the output signal at -10 volts.

Between time one (Tl) and time two (T2), input signal number one goes in a negative direction but input signal number two continues to go positive. Now the difference between the input signals decreases

rapidly from +1 volt. Halfway between Tl and T2 (the dotted vertical line), input signal number one and input signal number two are equal in amplitude. The difference between the input signals is 0 volts and this causes the output signal to be 0 volts. From this point to T2 the difference between the input signals is a negative value. At T2:

Bias = (input signal #1) - (input signal #2) Bias = (OV) - (+1V) Bias = +1V

From time two (T2) to time three (T3), input signal number one goes negative and input signal number two goes to zero. The difference between them stays constant at -1 volt. Therefore, the output signal stays at a +10-volt level for the entire time period from T2 to T3. At T3 the bias condition will be:

Bias = (input signal #1) - (input signal #2) Bias = (OV)- (-1V) Bias = +1V

Between T3 and T4 input signal number one goes to zero while input signal number two goes negative. This, again, causes a rapid change in the difference between the input signals. Halfway between T3 and T4 (the dotted vertical line) the two input signals are equal in amplitude; therefore, the difference between the input signals is 0 volts, and the output signal becomes 0 volts. From that point to T4, the difference between the input signals becomes a positive voltage. At T4:

Bias = (input signal #1) - (input signal #2) Bias = (OV)- (-1V) Bias = +1V

(The sequence of events from T4 to T8 are the same as those of TO to T4.)

As you have seen, this amplifier amplifies the difference between two input signals. But this is NOT a differential amplifier. A differential amplifier has two inputs and two outputs. The circuit you have just been shown has only one output. Well then, how does a differential amplifier schematic look?

TYPICAL DIFFERENTIAL AMPLIFIER CIRCUIT

Figure 3-6 is the schematic diagram of a typical differential amplifier. Notice that there are two inputs and two outputs. This circuit requires two transistors to provide the two inputs and two outputs. If you look at one input and the transistor with which it is associated, you will see that each transistor is a common-emitter amplifier for that input (input one and Ql; input two and Q2). Rl develops the signal at input one for Ql, and R5 develops the signal at input two for Q2. R3 is the emitter resistor for both Ql and Q2. Notice that R3 is NOT bypassed. This means that when a signal at input one affects the current through Ql, that signal is developed by R3. (The current through Ql must flow through R3; as this current changes, the voltage developed across R3 changes.) When a signal is developed by R3, it is applied to the emitter of Q2. In the same way, signals at input two affect the current of Q2, are developed by R3, and are felt on the emitter of Ql. R2 develops the signal for output one, and R4 develops the signal for output two.

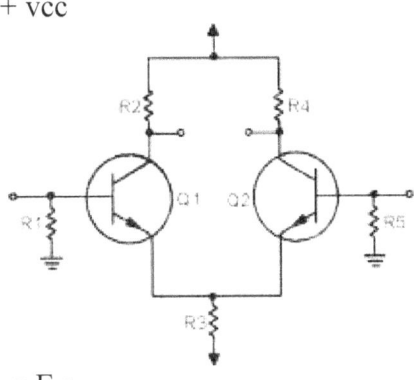

Figure 3-6.—Differential amplifier.

Even though this circuit is designed to have two inputs and two outputs, it is not necessary to use both inputs and both outputs. (Remember, a differential amplifier was defined as having two possible inputs and two possible outputs.) A differential amplifier can be connected as a single-input, single-output device; a single-input, differential-output device; or a differential-input, differential-output device.

Q-1. How many inputs and outputs are possible with a differential amplifier?

Q-2. What two transistor amplifier configurations are combined in the single-transistor, two-input, single-output difference amplifier?

Q-3. If the two input signals of a difference amplifier are in phase and equal in amplitude, what will the output signal be?

Q-4. If the two input signals to a difference amplifier are equal in amplitude and 180 degrees out of phase, what will the output signal be?

Q-5. If only one input signal is used with a difference amplifier, what will the output signal be?

Q-6. If the two input signals to a difference amplifier are equal in amplitude but neither in phase nor 180 degrees out of phase, what will the output signal be?

SINGLE-INPUT, SINGLE-OUTPUT, DIFFERENTIAL AMPLIFIER

Figure 3-7 shows a differential amplifier with one input (the base of Q1) and one output (the collector of Q2). The second input (the base of Q2) is grounded and the second output (the collector of Q1) is not used.

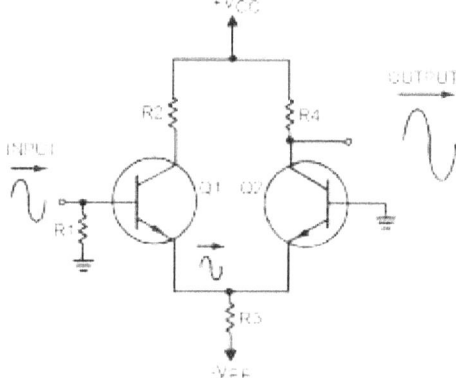

Figure 3-7.—Single-input, single-output differential amplifier.

When the input signal developed by R1 goes positive, the current through Q1 increases. This increased current causes a positive-going signal at the top of R3. This signal is felt on the emitter of Q2. Since the base of Q2 is grounded, the current through Q2 decreases with a positive-going signal on the emitter. This decreased current causes less voltage drop across R4. Therefore, the voltage at the bottom of R4 increases and a positive-going signal is felt at the output.

When the input signal developed by R1 goes negative, the current through Q1 decreases. This decreased current causes a negative-going signal at the top of R3. This signal is felt on the emitter of Q2. When the emitter of Q2 goes negative, the current through Q2 increases. This increased current causes more of a voltage drop across R4. Therefore, the voltage at the bottom of R4 decreases and a negative-going signal is felt at the output.

This single-input, single-output, differential amplifier is very similar to a single-transistor amplifier as far as input and output signals are concerned. This use of a differential amplifier does provide amplification of a.c. or d.c. signals but does not take full advantage of the characteristics of a differential amplifier.

SINGLE-INPUT, DIFFERENTIAL-OUTPUT, DIFFERENTIAL AMPLIFIER

In chapter one of this module you were shown several phase splitters. You should remember that a phase splitter provides two outputs from a single input. These two outputs are 180 degrees out of phase with each other. The single-input, differential-output, differential amplifier will do the same thing.

Figure 3-8 shows a differential amplifier with one input (the base of Q1) and two outputs (the collectors of Q1 and Q2). One output is in phase with the input signal, and the other output is 180 degrees out of phase with the input signal. The outputs are differential outputs.

COMBINED DIFFERENTIAL OUTPUT

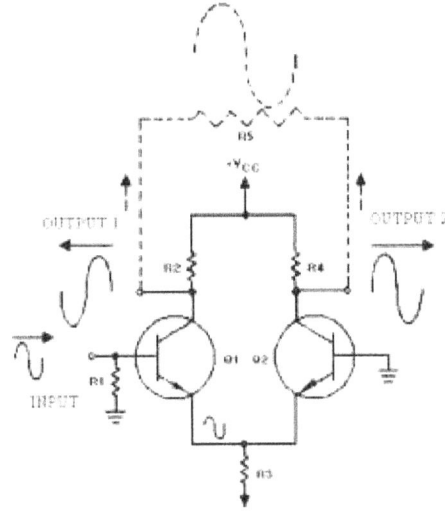

Figure 3-8.—Single-input, differential-output differential amplifier.

This circuit's operation is the same as for the single-input, single-output differential amplifier just described. However, another output is obtained from the bottom of R2. As the input signal goes positive, thus causing increased current through Q1, R2 has a greater voltage drop. The output signal at the bottom of R2 therefore is negative going. A negative-going input signal will decrease current and reverse the polarities of both output signals.

Now you see how a differential amplifier can produce two amplified, differential output signals from a single-input signal. One further point of interest about this configuration is that if a combined output signal is taken between outputs number one and two, this single output will be twice the amplitude of the individual outputs. In other words, you can double the gain of the differential amplifier (single output) by taking the output signal between the two output terminals. This single-output signal will be in phase with the input signal. This is shown by the phantom signal above R5 (the phantom resistor connected between outputs number one and two would be used to develop this signal).

DIFFERENTIAL-INPUT, DIFFERENTIAL-OUTPUT, DIFFERENTIAL AMPLIFIER

When a differential amplifier is connected with a differential input and a differential output, the full potential of the circuit is used. Figure 3-9 shows a differential amplifier with this type of configuration (differential-input, differential-output).

COMBINED OUTPUT

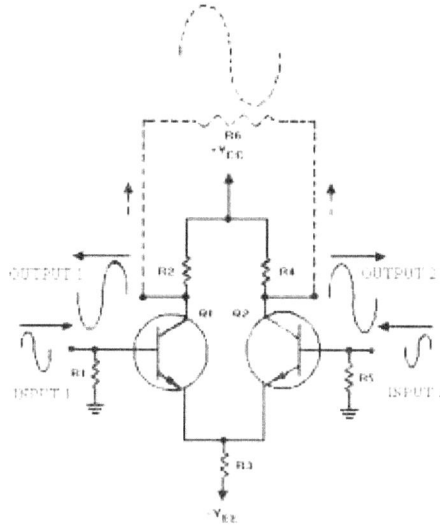

Figure 3-9.—Differential-input, differential-output differential amplifier.

Normally, this configuration uses two input signals that are 180 degrees out of phase. This causes the difference (differential) signal to be twice as large as either input alone. (This is just like the two-input, single-output difference amplifier with input signals that are 180 degrees out of phase.)

Output number one is a signal that is in phase with input number two, and output number two is a signal that is in phase with input number one. The amplitude of each output signal is the input signal multiplied by the gain of the amplifier. With 180-degree-out-of-phase input signals, each output signal is greater in amplitude than either input signal by a factor of the gain of the amplifier.

When an output signal is taken between the two output terminals of the amplifier (as shown by the phantom connections, resistor, and signal), the combined output signal is twice as great in amplitude as either signal at output number one or output number two. (This is because output number one and output number two are 180 degrees out of phase with each other.) When the input signals are 180 degrees out of phase, the amplitude of the combined output signal is equal to the amplitude of one input signal multiplied by two times the gain of the amplifier.

When the input signals are not 180 degrees out of phase, the combined output signal taken across output one and output two is similar to the output that you were shown for the two-input, single-output, difference amplifier. The differential amplifier can have two outputs (180 degrees out of phase with each other), or the outputs can be combined as shown in figure 3-9.

In answering Q7 through Q9 use the following information: All input signals are sine waves with a peak-to-peak amplitude of 10 millivolts. The gain of the differential amplifier is 10.

Q-7. If the differential amplifier is configured with a single input and a single output, what will the peak-to-peak amplitude of the output signal be?

Q-8. If the differential amplifier is configured with a single input and differential outputs, what will the output signals be?

Q-9. If the single-input, differential-output, differential amplifier has an output signal taken between the two output terminals, what will the peak-to-peak amplitude of this combined output be?

In answering Q10 through Q14 use the following information: A differential amplifier is configured with a differential input and a differential output. All input signals are sine waves with a peak-to-peak amplitude of 10 millivolts. The gain of the differential amplifier is 10.

Q-10. If the input signals are in phase, what will be the peak-to-peak amplitude of the output signals?

Q-ll. If the input signals are 180 degrees out of phase with each other, what will be the peak-to-peak amplitude of the output signals?

Q-12. If the input signals are 180 degrees out of phase with each other, what will the phase relationship be between (a) the output signals and (b) the input and output signals?

Q-13. If the input signals are 180 degrees out of phase with each other and a combined output is taken between the two output terminals, what will the amplitude of the combined output signal be?

Q-14. If the input signals are 90 degrees out of phase with each other and a combined output is taken between the two output terminals, (a) what will the peak-to-peak amplitude of the combined output signal be, and (b) will the combined output signal be a sine wave?

OPERATIONAL AMPLIFIERS

An OPERATIONAL AMPLIFIER (OP AMP) is an amplifier which is designed to be used with other circuit components to perform either computing functions (addition, subtraction) or some type of transfer operation, such as filtering. Operational amplifiers are usually high-gain amplifiers with the amount of gain determined by feedback.

Operational amplifiers have been in use for some time. They were originally developed for analog (non-digital) computers and used to perform mathematical functions. Operational amplifiers were not used in other devices very much because they were expensive and more complicated than other circuits.

Today many devices use operational amplifiers. Operational amplifiers are used as d.c. amplifiers, a.c. amplifiers, comparators, oscillators (which are covered in NEETS, Module 9), filter circuits, and many other applications. The reason for this widespread use of the operational amplifier is that it is a very versatile and efficient device. As an integrated circuit (chip) the operational amplifier has become an inexpensive and readily available "building block" for many devices. In fact, an operational amplifier in integrated circuit form is no more expensive than a good transistor.

CHARACTERISTICS OF AN OPERATIONAL AMPLIFIER

The schematic symbols for an operational amplifier are shown in figure 3-10. View (A) shows the power supply requirements while view (B) shows only the input and output terminals. An operational amplifier is a special type of high-gain, d.c. amplifier. To be classified as an operational amplifier, the circuit must have certain characteristics. The three most important characteristics of an operational amplifier are:

    1. Very high gain
    2. Very high input impedance
    3. Very low output impedance

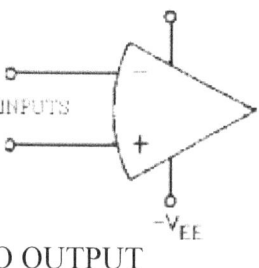

O OUTPUT
(A)
Figure 3-10A.—Schematic symbols of an operational amplifier.

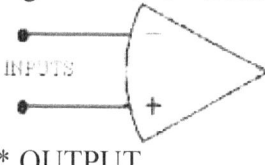

* OUTPUT

Figure 3-10B.—Schematic symbols of an operational amplifier.

Since no single amplifier stage can provide all these characteristics well enough to be considered an operational amplifier, various amplifier stages are connected together. The total circuit made up of these individual stages is called an operational amplifier. This circuit (the operational amplifier) can be made up of individual components (transistors, resistors, capacitors, etc.), but the most common form of the operational amplifier is an integrated circuit. The integrated circuit (chip) will contain the various stages of the operational amplifier and can be treated and used as if it were a single stage.

BLOCK DIAGRAM OF AN OPERATIONAL AMPLIFIER

Figure 3-11 is a block diagram of an operational amplifier. Notice that there are three stages within the operational amplifier.

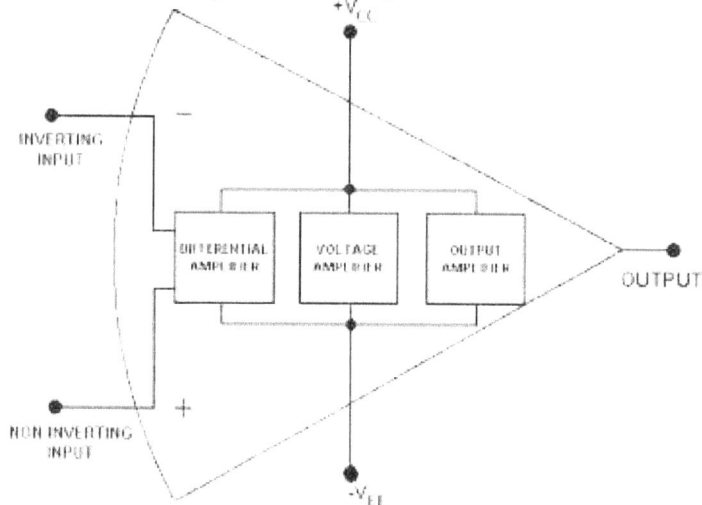

Figure 3-11.—Block diagram of an operational amplifier.

The input stage is a differential amplifier. The differential amplifier used as an input stage provides differential inputs and a frequency response down to d.c. Special techniques are used to provide the high input impedance necessary for the operational amplifier.

The second stage is a high-gain voltage amplifier. This stage may be made from several transistors to provide high gain. A typical operational amplifier could have a voltage gain of 200,000. Most of this gain comes from the voltage amplifier stage.

The final stage of the OP AMP is an output amplifier. The output amplifier provides low output impedance. The actual circuit used could be an emitter follower. The output stage should allow the operational amplifier to deliver several milliamperes to a load.

Notice that the operational amplifier has a positive power supply (+V C c) and a negative power supply (-V EE). This arrangement enables the operational amplifier to produce either a positive or a negative output.

The two input terminals are labeled "inverting input" (-) and "noninverting input" (+). The operational amplifier can be used with three different input conditions (modes). With differential inputs (first mode), both input terminals are used and two input signals which are 180 degrees out of phase with each other are used. This produces an output signal that is in phase with the signal on the noninverting input. If the noninverting input is grounded and a signal is applied to the inverting input (second mode), the output signal will be 180 degrees out of phase with the input signal (and one-half the amplitude of the first mode output). If the inverting input is grounded and a signal is applied to the noninverting input (third mode), the output signal will be in phase with the input signal (and one-half the amplitude of the first mode output).

Q-15. What are the three requirements for an operational amplifier?

Q-16. What is the most commonly used form of the operational amplifier?

Q-l 7. Draw the schematic symbol for an operational amplifier.

Q-18. Label the parts of the operational amplifier shown in figure 3-12.

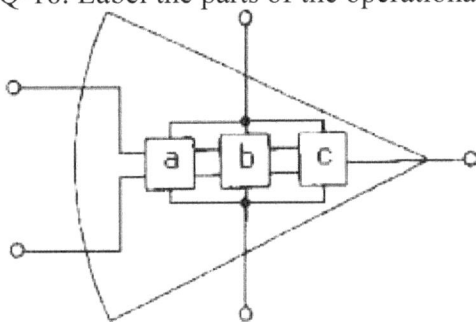

Figure 3-12.—Operational amplifier.

CLOSED-LOOP OPERATION OF AN OPERATIONAL AMPLIFIER

Operational amplifiers can have either a closed-loop operation or an open-loop operation. The operation (closed-loop or open-loop) is determined by whether or not feedback is used. Without feedback the operational amplifier has an open-loop operation. This open-loop operation is practical only when the operational amplifier is used as a comparator (a circuit which compares two input signals or compares an input signal to some fixed level of voltage). As an amplifier, the open-loop operation is not practical because the very high gain of the operational amplifier creates poor stability. (Noise and other unwanted signals are amplified so much in open-loop operation that the operational amplifier is usually not used in this way.) Therefore, most operational amplifiers are used with feedback (closed-loop operation).

Operational amplifiers are used with degenerative (or negative) feedback which reduces the gain of the operational amplifier but greatly increases the stability of the circuit. In the closed-loop configuration, the output signal is applied back to one of the input terminals. This feedback is always degenerative (negative). In other words, the feedback signal always opposes the effects of the original input signal. One result of degenerative feedback is that the inverting and noninverting inputs to the operational amplifier will be kept at the same potential.

Closed-loop circuits can be of the inverting configuration or noninverting configuration. Since the inverting configuration is used more often than the noninverting configuration, the inverting configuration will be shown first.

Inverting Configuration

Figure 3-13 shows an operational amplifier in a closed-loop, inverting configuration. Resistor R2 is used to feed part of the output signal back to the input of the operational amplifier.

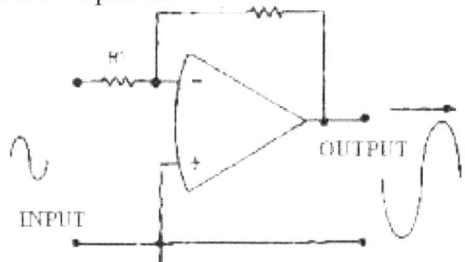

Figure 3-13.—Inverting configuration.

At this point it is important to keep in mind the difference between the entire circuit (or operational circuit) and the operational amplifier. The operational amplifier is represented by the triangle-like symbol while the operational circuit includes the resistors and any other components as well as the operational amplifier. In other words, the input to the circuit is shown in figure 3-13, but the signal at the inverting input of the operational amplifier is determined by the feedback signal as well as by the circuit input signal.

As you can see in figure 3-13, the output signal is 180 degrees out of phase with the input signal. The feedback signal is a portion of the output signal and, therefore, also 180 degrees out of phase with the input signal. Whenever the input signal goes positive, the output signal and the feedback signal go negative. The result of this is that the inverting input to the operational amplifier is always very close to 0 volts with this configuration. In fact, with the noninverting input grounded, the voltage at the inverting input to the operational amplifier is so small compared to other voltages in the circuit that it is considered to be VIRTUAL GROUND. (Remember, in a closed-loop operation the inverting and noninverting inputs are at the same potential.)

Virtual ground is a point in a circuit which is at ground potential (0 volts) but is NOT connected to ground. Figure 3-14, (view A) (view B) and (view C), shows an example of several circuits with points at virtual ground.

R1 -Wv-
1Q
POINT
(A)
R2 -wv-
1Q
+10V V1 —
■10V 1=1 V2
z
(A)

Figure 3-14A.—Virtual ground circuits.

POINT
(A)

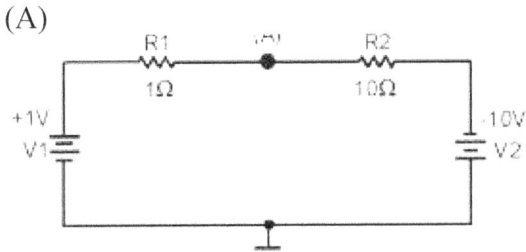

(B)

Figure 3-14B.—Virtual ground circuits.

R1
POINT
(A)
R2
•5V VI
-IV - V2
(C)

Figure 3-14C.—Virtual ground circuits.

In view (A), VI (the left-hand battery) supplies +10 volts to the circuit while V2 (the right-hand battery) supplies -10 volts to the circuit. The total difference in potential in the circuit is 20 volts.

The total resistance of the circuit can be calculated:
R T = Rl+ R 2 R T =1Q + 1Q R T = 2Q

Now that the total resistance is known, the circuit current can be calculated:
The voltage drop across Rl can be computed:
Eri = Ri* It Em = 1Q x 10A Em = 10V

The voltage at point A would be equal to the voltage of VI minus the voltage drop of Rl.

Voltage at point A = VI - E Ri Voltage at point A = +10V- 10V Voltage at point A = OV

To check this result, compute the voltage drop across R2 and subtract this from the voltage at point A. The result should be the voltage of V2.

Er2 = Ri* It Er 2 = 1Q x 10A Ers = 10V
V2 = (voltage at point A) - (Erz) V2 = (OV)-(10V) V2 = -10V

It is not necessary that the voltage supplies be equal to create a point of virtual ground. In view (B) VI supplies +1 volt to the circuit while V2 supplies -10 volts. The total difference in potential is 11 volts. The total resistance of this circuit (Rl + R2) is 11 ohms. The total current (I T) is 1 ampere. The voltage drop across Rl (E R1 = Ri x I T) is 1 volt. The voltage drop across R2 (E R2 = R2 x I T) is 10 volts. The voltage at point A can be computed:

Voltage at point A = VI - Eri Voltage at point A = (+1V) - (+ 1V) Voltage at point A = OV

So point A is at virtual ground in this circuit also. To check the results, compute the voltage at V2.

V2 = (voltage at point A) - Eri V2 = (OV)-(+10V) V2 = -10V

You can compute the values for view (C) and prove that point A in that circuit is also at virtual ground.

The whole point is that the inverting input to the operational amplifier shown in figure 3-13 is at virtual ground since it is at 0 volts (for all practical purposes). Because the inverting input is at 0 volts, there will be no current (for all practical purposes) flowing into the operational amplifier from the connection point of Rl and R2.

Given these conditions, the characteristics of this circuit are determined almost entirely by the values of Rl and R2. Figure 3-15 should help show how the values of Rl and R2 determine the circuit characteristics.

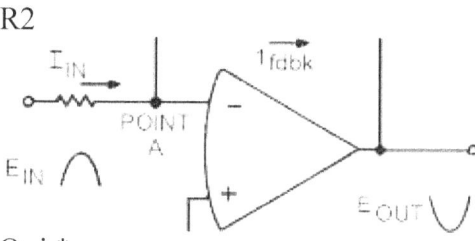

O- i *

Figure 3-15.—Current flow in the operational circuit.

NOTE: It should be stressed at this point that for purpose of explanation the operational amplifier is a theoretically perfect amplifier. In actual practice we are dealing with less than perfect. In the practical operational amplifier there will be a slight input current with a resultant power loss. This small signal can be measured at the theoretical point of virtual ground. This does not indicate faulty operation.

The input signal causes current to flow through Rl. (Only the positive half cycle of the input signal is shown and will be discussed.) Since the voltage at the inverting input of the operational amplifier is at 0 volts, the input current (I in) is computed by:

E-
T = m
m r7

The output signal (which is opposite in phase to the input signal) causes a feedback current (Ifdbk) to flow through R2. The left-hand side of R2 is at 0 volts (point A) and the right-hand side is at E out . Therefore, the feedback current is computed by:

T _ ~ E 0Ut
l mv ~ —
K 2

(The minus sign indicates that E out is 180 degrees out of phase with E in and should not be confused with output polarity.)

Since no current flows into or out of the inverting input of the operational amplifier, any current reaching point A from Rl must flow out of point A through R2. Therefore, the input current (I in) and the feedback current (I fd bk) must be equal. Now we can develop a mathematical relationship between the input and output signals and Rl and R2.

Mathematically:
By sub stituition:
F —F
in _ out
If you multiply both sides of the equation by Rl:

$$E_{in} = -\left(\frac{E_{out}}{R2}\right)(R1)$$

If you divide both sides of the equation by E out:

$$\frac{E_{in}}{E_{out}} = -\frac{R1}{R2}$$

By inverting both sides of the equation:

$$\frac{E_{in}}{E_{out}} = -\frac{R2}{R1}$$

You should recall that the voltage gain of a stage is defined as the output voltage divided by the input voltage:

$$Gain = E_{out} / E_{in}$$

Therefore, the voltage gain of the inverting configuration of the operational amplifier is expressed by the equation:

$$h,$$

(As stated earlier, the minus sign indicates that the output signal is 180 degrees out of phase with the input signal.)

Noninverting Configuration

Figure 3-16 shows a noninverting configuration using an operational amplifier. The input signal (E in) is applied directly to the noninverting (+) input of the operational amplifier. Feedback is provided by

coupling part of the output signal (E out) back to the inverting (-) input of the operational amplifier. Rl and R2 act as voltage divider that allows only a part of the output signal to be applied as feedback (E fdbk).

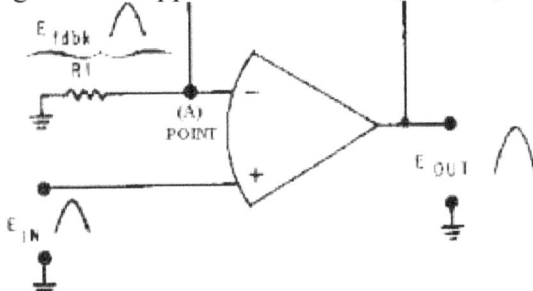

Figure 3-16.—Noninverting configuration.

Notice that the input signal, output signal, and feedback signal are all in phase. (Only the positive alternation of the signal is shown.) It may appear as if the feedback is regenerative (positive) because the feedback and input signals are in phase. The feedback is, in reality, degenerative (negative) because the input signals is applied to the noninverting input and the feedback signal is applied to the inverting input, (Remember, that the operational amplifier will react to the difference between the two inputs.)

Just as in the inverting configuration, the feedback signal is equal to the input signal (for all practical purposes). This time, however, the feedback signal is in phase with the input signal.

Therefore:

E m = E fdbli

Given this condition, you can calculate the gain of the stage in terms of the resistors (Rl and R2). The gain of the stage is defined as:

Gain =

$$\frac{E_{out}}{E_{in}}$$

Since: $E_{in} = E_{fdbk}$

Then: $Gain = \frac{E_{out}}{E_{fdbk}}$

The feedback signal (E_{fdbk}) can be shown in terms of the output signal (E_{out}) and the voltage divider (R1 and R2). The voltage divider has the output signal on one end and ground (0 volts) on the other end. The feedback signal is that part of the output signal developed by R1 (at point A). Another way to look at it is that the feedback signal is the amount of output signal left (at point A) after part of the output signal

has been dropped by R2. In either case, the feedback signal (E_{fdbk}) is the ratio of R1 to the entire voltage divider (R1 + R2) multiplied by the output signal (E_{out}).

Mathematically, the relationship of the output signal, feedback signal, and voltage divider is:

$$E_{fdbk} = \frac{R1}{R1+R2}(E_{out})$$

If you divide both sides of the equation by E_{out}:

$$\frac{E_{fdbk}}{E_{out}} = \frac{R1}{R1+R2}$$

By inverting both sides of the equation:

$$\frac{E_{out}}{E_{fdbk}} = \frac{R1+R2}{R1}$$

Separating the right-hand side:

$$\frac{E_{out}}{E_{fdbk}} = \frac{R1}{R1} + \frac{R2}{R1}$$

Remember:

$$Gain = \frac{E_{out}}{E_{fdbk}}$$

Therefore, by substitution:

$$Gain = \frac{R2}{R1} + 1$$

You can now see that the gain of the noninverting configuration is determined by the resistors. The formula is different from the one used for the inverting configuration, but the gain is still determined by the values of R1 and R2.

BANDWIDTH LIMITATIONS

As with most amplifiers, the gain of an operational amplifier varies with frequency. The specification sheets for operational amplifiers will usually state the open-loop (no feedback) gain for d.c. (or 0 hertz). At higher frequencies, the gain is much lower. In fact, for an operational amplifier, the gain decreases quite rapidly as frequency increases.

Figure 3-17 shows the open-loop (no feedback) frequency-response curve for a typical operational amplifier. As you should remember, bandwidth is measured to the half-power points of a frequency-response curve. The frequency-response curve shows that the bandwidth is only 10 hertz with this

configuration. The UNITY GAIN POINT, where the signal out will have the same amplitude as the signal in (the point at which the gain of the amplifier is 1), is 1 megahertz for the amplifier. As you can see, the frequency response of this amplifier drops off quite rapidly.

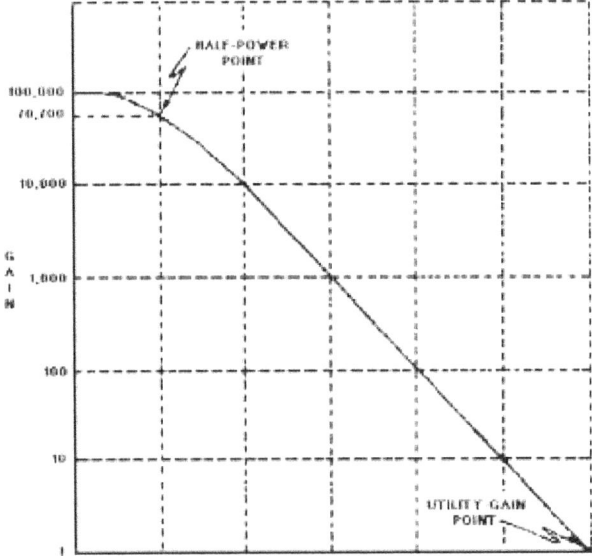

Figure 3-17.—Open-loop frequency-response curve.

Figure 3-17 is the open-loop frequency-response curve. You have been told that most operational amplifiers are used in a closed-loop configuration. When you look at the frequency-response curve for a closed-loop configuration, one of the most interesting and important aspects of the operational amplifier becomes apparent: The use of degenerative feedback increases the bandwidth of an operational amplifier circuit.

This phenomenon is another example of the difference between the operational amplifier itself and the operational-amplifier circuit (which includes the components in addition to the operational amplifier). You should also be able to see that the external resistors not only affect the gain of the circuit, but the bandwidth as well.

You might wonder exactly how the gain and bandwidth of a closed-loop, operational-amplifier circuit are related. Figure 3-18 should help to show you the relationship. The frequency-response curve shown in figure 3-18 is for a circuit in which degenerative feedback has been used to decrease the circuit gain to 100 (from 100,000 for the operational amplifier). Notice that the half-power point of this curve is just slightly above 10 kilohertz.

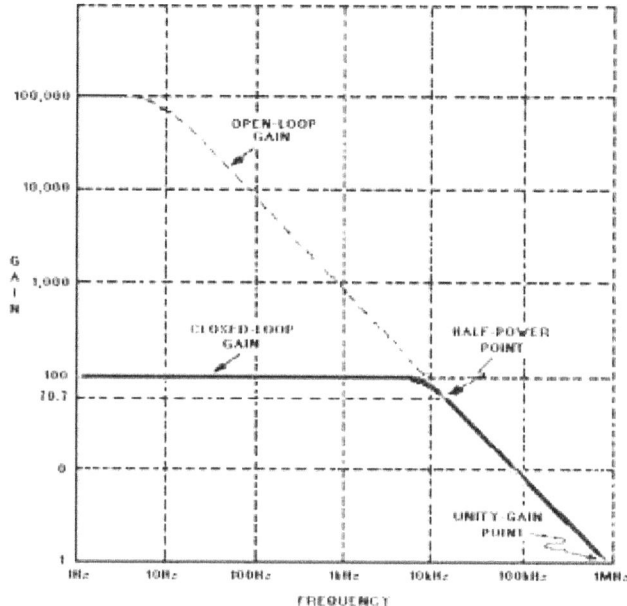

Figure 3-18.—Closed-loop frequency-response curve for gain of 100.

Now look at figure 3-19. In this case, more feedback has been used to decrease the gain of the circuit to 10. Now the bandwidth of the circuit is extended to about 100 kilohertz.

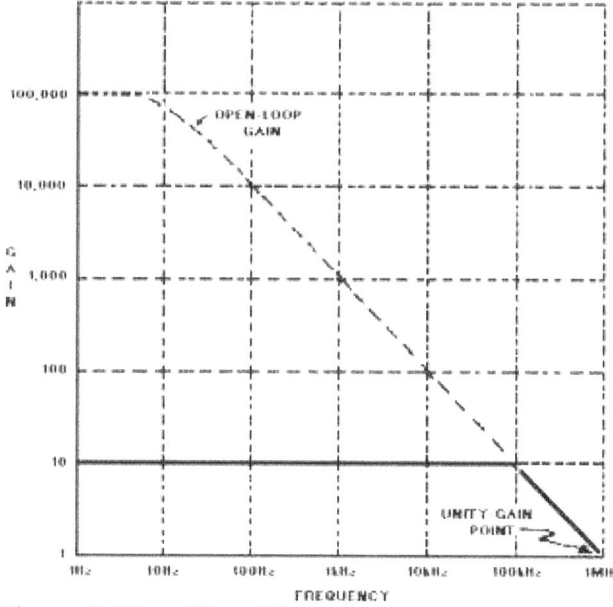

Figure 3-19.—Closed-loop frequency-response curve for gain of 10.

The relationship between circuit gain and bandwidth in an operational-amplifier circuit can be expressed by the GAIN-BANDWIDTH PRODUCT (GAIN x BANDWIDTH = UNITY GAIN POINT). In other words, for operational-amplifier circuits, the gain times the bandwidth for one configuration of an operational amplifier will equal the gain times the bandwidth for any other configuration of the same operational amplifier. In other words, when the gain of an operational-amplifier circuit is changed (by changing the value of feedback or input resistors), the bandwidth also changes. But the gain times the bandwidth of the first configuration will equal the gain times the bandwidth of the second configuration. The following example should help you to understand this concept.

The frequency-response curves shown in figures 3-17, 3-18, and 3-19 have a gain-bandwidth product of 1,000,000. In figure 3-17, the gain is 100,000 and the bandwidth is 10 hertz. The gain-bandwidth product is 100,000 times 10 (Hz), or 1,000,000. In figure 3-18, the gain has been reduced to 100 and the bandwidth increases to 10 kilohertz. The gain-bandwidth product is 100 times 10,000 (Hz) which is also equal to 1,000,000. In figure 3-19 the gain has been reduced to 10 and the bandwidth is 100 kilohertz. The gain-bandwidth product is 10 times 100,000 (Hz), which is 1,000,000. If the gain were reduced to 1, the bandwidth would be 1 megahertz (which is shown on the frequency-response curve as the unity-gain point) and the gain-bandwidth product would still be 1,000,000.

Q-19. What does the term "closed-loop" mean in the closed-loop configuration of an operational amplifier?

In answering Q20, Q21, and Q23, select the correct response from the choices given in the parentheses.

Q-20. In a closed-loop configuration the output signal is determined by (the input signal, the feedback signal, both).

Q-21. In the inverting configuration, the input signal is applied to the (a) (inverting, noninverting) input and the feedback signal is applied to the (b) (inverting, noninverting) input.

Q-22. In the inverting configuration, what is the voltage (for all practical purposes) at the inverting input to the operational amplifier if the input signal is a 1-volt, peak-to-peak sine wave?

Q-23. In the inverting configuration when the noninverting input is grounded, the inverting input is at (signal, virtual) ground.

Q-24. In a circuit such as that shown in figure 3-15, ifRl has a value of 100 ohms and R2 has a value of 1 kilohm and the input signal is at a value of+ 5 millivolts, what is the value of the output signal?

Q-25. If the unity-gain point of the operational amplifier used in question 24 is 500 kilohertz, what is the bandwidth of the circuit?

Q-26. In a circuit such as that shown in figure 3-16, ifRl has a value of 50 ohms and R2 has a value of 250 ohms and the input signal has a value of +10 millivolts, what is the value of the output signal?

Q-27. If the open-loop gain of the operational amplifier used in question 26 is 200,000 and the open-loop bandwidth is 30 hertz, what is the closed loop bandwidth of the circuit?

APPLICATIONS OF OPERATIONAL AMPLIFIERS

Operational amplifiers are used in so many different ways that it is not possible to describe all of the applications. Entire books have been written on the subject of operational amplifiers. Some books are devoted entirely to the applications of operational amplifiers and are not concerned with the theory of operation or other circuits at all. This module, as introductory material on operational amplifiers, will show you only two common applications of the operational amplifier: the summing amplifier and the difference amplifier. For ease of explanation the circuits shown for these applications will be explained with d.c. inputs and outputs, but the circuit will work as well with a.c. signals.

Summing Amplifier (Adder)

Figure 3-20 is the schematic of a two-input adder which uses an operational amplifier. The output level is determined by adding the input signals together (although the output signal will be of opposite polarity compared to the sum of the input signals).

R1 1kO
R3 1kO
E1

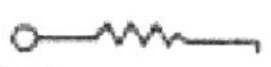

R2
E2 O WNr
E
IN

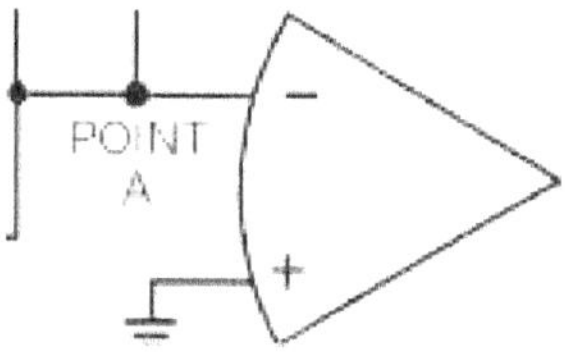

POINT A

E OUT,
1

Figure 3-20.—Two-input adder.

If the signal on input number one (E1) is +3 volts and the signal on input number two (E2) is +4 volts, the output signal (E out) should be -7 volts [(+3 V) + (+4 V) = +7 V and change the polarity to get -7 V].

With +3 volts at E1 and 0 volts at point A (which is at virtual ground), the current through R1 must be 3 milliamperes.

Mathematically:

(The + sign indicates a current flow from right to left.)

By the same sort of calculation, with +4 volts at E2 and 0 volts at point A the current through R2 must be 4 milliamps.

This means that a total of 7 milliamps is flowing from point A through R1 and R2. If 7 milliamps is flowing from point A, then 7 milliamps must be flowing into point A. The 7 milliamps flowing into point A flows through R3 causing 7 volts to be developed across R3. With point A at 0 volts and 7 volts developed across R3, the voltage potential at E out must be a -7 volts. Figure 3-21 shows these voltages and currents.

$I_{R1} = \dfrac{E_1}{R_1}$

+3V
+3mA
M
R3

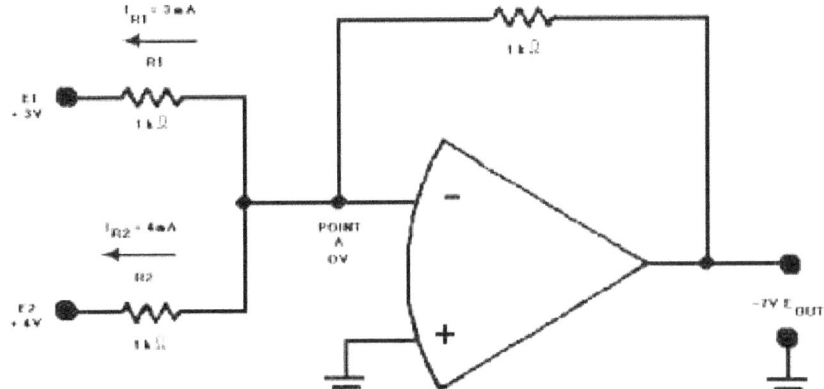

Figure 3-21.—Current and voltage in a two-input adder.

An adder circuit is not restricted to two inputs. By adding resistors in parallel to the input terminals, any number of inputs can be used. The adder circuit will always produce an output that is equal to the sum of the input signals but opposite in polarity. Figure 3-22 shows a five-input adder circuit with voltages and currents indicated.

El
R1 1 k ii

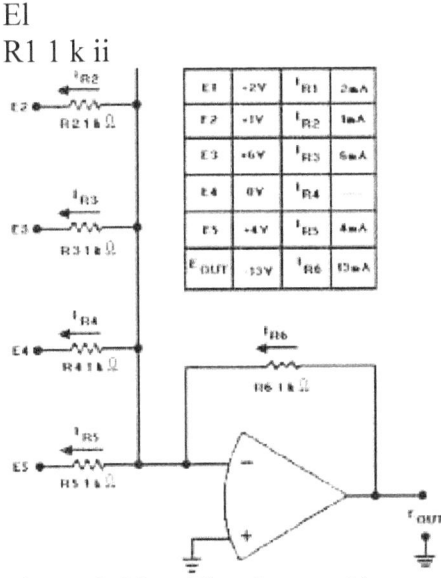

Figure 3-22.—Five-input adder.

The previous circuits have been adders, but there are other types of summing amplifiers. A summing amplifier can be designed to amplify the results of adding the input signals. This type of circuit actually multiplies the sum of the inputs by the gain of the circuit.

Mathematically (for a three-input circuit):
Eom" gain (El + E2 + E3)
If the circuit gain is -10:
E^-10 (El + E2 + E3)

The gain of the circuit is determined by the ratio between the feedback resistor and the input resistors. To change figure 3-20 to a summing amplifier with a gain of -10, you would replace the feedback resistor (R3) with a 10-kilohm resistor. This new circuit is shown in figure 3-23.

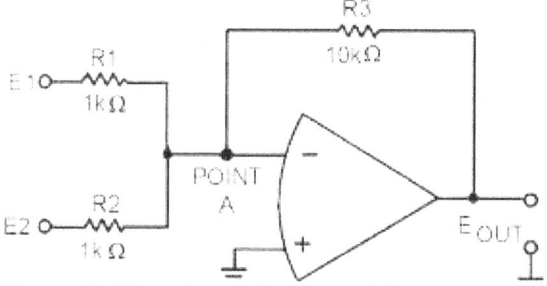

Figure 3-23.—Summing amplifier.

If this circuit is designed correctly and the input voltages (E1 and E2) are +2 volts and +3 volts, respectively, the output voltage (E out) should be:

Eort-gain (E1 + E2) E rat =-10[C+2V) + (+ 3V)] E™t="10 (+5) E™,= -50V

To see if this output (-50 V) is what the circuit will produce with the inputs given above, start by calculating the currents through the input resistors, R1 and R2 (remember that point A is at virtual ground):

I E1
I 2V I ri = 2rnA I E2
i -3V
I R2 =3rnA

Next, calculate the current through the feedback resistor (R3):

*R3 (IrI + ^R2)
I R3 = -(2mA + 3 mA)
1^3 = ~5 mA (The minus sign indicates current flow from left to right.)

Finally, calculate the voltage dropped across R3 (which must equal the output voltage):

E out =(I R3 x R3)
E out =(-5 mA x 10 kfi)
E out =-50V

As you can see, this circuit performs the function of adding the inputs together and multiplying the result by the gain of the circuit.

One final type of summing amplifier is the SCALING AMPLIFIER. This circuit multiplies each input by a factor (the factor is determined by circuit design) and then adds these values together. The factor that is used to multiply each input is determined by the ratio of the feedback resistor to the input resistor. For example, you could design a circuit that would produce the following output from three inputs (E1, E2, E3):

-[(2 xE1) + (4 x E2) + (3x E3)]

Using input resistors R1 for input number one (E1), R2 for input number two (E2), R3 for input number three (E3), and R4 for the feedback resistor, you could calculate the values for the resistors:

2-*i
Ri

Any resistors that will provide the ratios shown above could be used. If the feedback resistor (R4) is a 12-kilohm resistor, the values of the other resistors would be:

Figure 3-24 is the schematic diagram of a scaling amplifier with the values calculated above.

R1 R?
E20—'WV

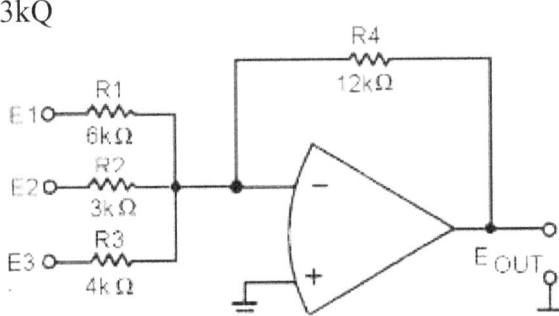

Figure 3-24.—Scaling amplifier.

To see if the circuit will produce the desired output, calculate the currents and voltages as done the previous circuits.

With:
E1 = +12V E2 = +3V E3 = +8V
the output should be:
E, ut = -[(2 x E1) + (4 x E2) + (3 x E3)]
E out = -[(2 x +12V) + (4 x +3V) + (3 x +SV)]
= -[(+ 24 V) + (+ 12 V) + (+ 24 V)]
E_ = -60 V

Calculate the current for each input:
I E1
T _ +12V IR1= W
I R1 = +2rnA
I +3V I E2 = +lmA
t _ E3
Ir3 _ R3
T _ +8V
I R3 =+2rnA
Ha ~ (^ri +1 R2 + hi) l u =-(2rnA + lmA + 2rnA)
l M = -5mA

Calculate the output voltage:
Eout = Er^
Eout = Ir-i x Ka
= (-5mA xl2kQ E™t = -60V

You have now seen how an operational amplifier can be used in a circuit as an adder, a summing amplifier, and a scaling amplifier.

Difference Amplifier (Subtractor)

A difference amplifier will produce an output based on the difference between the input signals. The subtractor circuit shown in figure 3-25 will produce the following output:

Eout = E2 - E1
R3 1kD

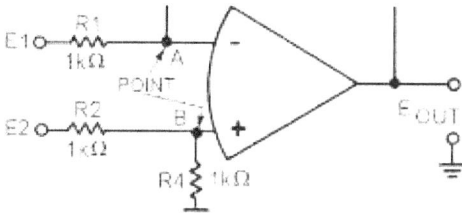

Figure 3-25.—Subtractor circuit.

Normally, difference amplifier circuits have the ratio of the inverting input resistor to the feedback resistor equal to the ratio of the noninverting input resistors. In other words, for figure 3-25:

Rl = R2 R3 R4

and, by inverting both sides:

R3 = R4 Rl R2

For ease of explanation, in the circuit shown in figure 3-25 all the resistors have a value of 1 kilohm, but any value could be used as long as the above ratio is true. For a subtractor circuit, the values of Rl and R3 must also be equal, and therefore, the values of R2 and R4 must be equal. It is NOT necessary that the value of Rl equal the value of R2.

Using figure 3-25, assume that the input signals are:

El = +3V E2 = +12V

The output signal should be:

E out = (+12V)-(+3V)

E out =+9V

To check this output, first compute the value of R2 plus R4:

R 2 + R4 = lkQ + lkQ R 2 + R^| = 2kQ

With this value, compute the current through R2 (I R2):

I - E2

R 2 +R4 + 12V

R2

2kQ I R2 =+6rnA

(indicating current flow from left to right)

Next, compute the voltage drop across R2 (E R2):

E E2 = R 2 x I E2

E E2 = lkQ x (+6rnA)

E R2 = +6V

Then compute the voltage at point B:

Then compute the voltage at point B:

Voltage at p oint B = E2 - E E2 Voltage at p oint B = (+ 12V) - (+6V) Voltage at p oint B = +6V

Since point B and point A will be at the same potential in an operational amplifier:

Voltage at point A = +6V

Now compute the voltage developed by Rl (E RI):

E R1 = (voltage at point A) - (El)

E R i= C+6V) - (+3V) E R1 = +3V

Compute the current through Rl (I R1):

3-38

= fjy. +3V

IR1 = lkQ I R1 = +3mA
Since: I R1 =I E3
Then: I R3 = +3mA
Compute the voltage developed by R3 (E R3):
ER3= (R3) x (IR3) E R3 = (IkO) x (+3mA) E R3 =+3V
Add this to the voltage at point A to compute the output voltage (E out):
Emit = (E E3) + (voltage at point A) Eout = C+3V) x (+6V) Eout = + 3V

As you can see, the circuit shown in figure 3-25 functions as a subtractor. But just as an adder is only one kind of summing amplifier, a subtractor is only one kind of difference amplifier. A difference amplifier can amplify the difference between two signals. For example, with two inputs (El and E2) and a gain of five, a difference amplifier will produce an output signal which is:

E™, = 5CE2-E1)

The difference amplifier that will produce that output is shown in figure 3-26. Notice that this circuit is the same as the subtractor shown in figure 3-25 except for the values of R3 and R4. The gain of this difference amplifier is:

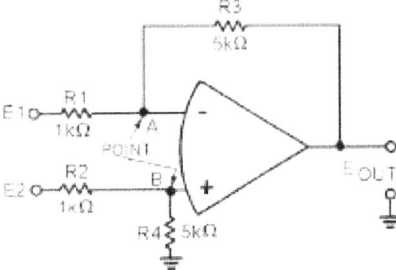

Figure 3-26.—Difference amplifier.

Then, for a difference amplifier:
Gain = -
5 Ida
Gain = —— 1 Ida
Gain = 5
R3 R4 Gam = — = — Rl R2

With the same inputs that were used for the subtractor, (El = + 3 V; E2 = + 12 V) the output of the difference amplifier should be five times the output of the subtractor (E out = + 45 V).

Following the same steps used for the subtractor:
First compute the value of R2 plus R4:
R2 + R4 = 1 kQ + 5 Ida R2 + R4 = 4 Ida With this value, compute the current through R2 (I R2):

E2
I
I
R2
E2
R 2 +R4 + 12V
6kQ I R2 =+2rnA

Next, compute the voltage drop across R2 (E R2):

$E_{R2} = (R2) \times (I_{R2})$ $E_{E2} = (1k\Omega) \times (+2mA)$ $E_{E2} = +2V$

Then, compute the voltage at point B:

Voltage at point B = E2 - Er2   Voltage at point B = (+12V) - (+2V)   Voltage at point B = +10V

Since point A and point B will be at the same potential in an operational amplifier:

Voltage at point A = +10V

Now compute the voltage developed by Rl (E_{R1}):

$E_{R1} =$

$E_{R1} =$

$E_{R1} =$

(voltage at point A) - (El) (+10V) - (+3V) + 7 V

Compute the current through Rl (I_{R1}):

I R1 = +7mA   Since: I R i = I R3   Then: I R3 = +7mA

Compute the voltage developed by R3 (E_{R3}):

Add this voltage to the voltage at point A to compute the output voltage (E_{out}):

This was the output desired, so the circuit works as a difference amplifier. Q-28. What is the difference between a summing amplifier and an adder circuit? Q-29. Can a summing amplifier have more than two inputs? Q-30. What is a scaling amplifier?

Refer to figure 3-27 in answering Q31 through Q33.

$E_{R3} = R3 \times I_{R3}$

$E_{E3} = (5k\Omega) \times (+7mA)$

+35V

$E_{out} = (E_{R3}) +$ (voltage at point A) $E_{out} = (+35V) + (+10V)$

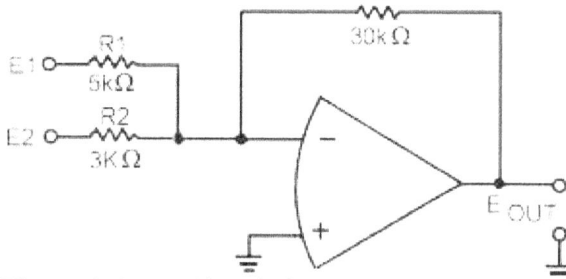

Figure 3-27.—Circuit for Q31 through Q33.

Q-31. What type of circuit is figure 3-27?

Q-32. If: El = +2V, and: E2 = +6V, then E out = ?

Q-33. What is the difference in potential between the inverting (-) and noninverting (+) inputs to the operational amplifier when: El = +6V, and E2 - +2V

Q-34. What is the difference between a subtractor and a difference amplifier?

Q-35. Can a difference amplifier have more than two inputs?

Refer to figure 3-28 in answering Q36 through Q38.

R3 -VW-

20kΩ

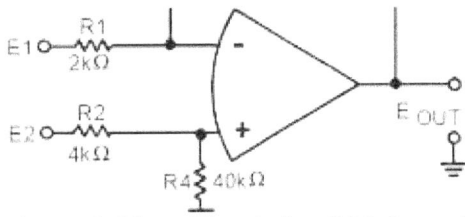

Figure 3-28.—Circuit for Q36 through Q38.

Q-36. What type of circuit is figure 3-28?

Q-37. If: E1 = +5V, and: E2 = +UV, then E OM = ?

Q-38. What is the difference in potential between the inverting (-) and noninverting (+) inputs to the operational amplifier when: E1 = +2V, and E2 = +4V

MAGNETIC AMPLIFIERS

You have now been shown various ways that electron tubes (NEETS, Module 6) and transistors (NEETS, Module 7) can be used to amplify signals. You have also been shown the way in which this is done. There is another type of amplifier in use—the MAGNETIC AMPLIFIER, sometimes called the MAG AMP.

The magnetic amplifier has certain advantages over other types of amplifiers. These include (1) high efficiency (up to 90 percent); (2) reliability (long life, freedom from maintenance, reduction of spare parts inventory); (3) ruggedness (shock and vibration resistance, high overload capability, freedom from effects of moisture); and (4) no warm-up time. The magnetic amplifier has no moving parts and can be hermetically sealed within a case similar to the conventional dry-type transformer.

However, the magnetic amplifier has a few disadvantages. For example, it cannot handle low-level signals; it is not useful at high frequencies; it has a time delay associated with the magnetic effects; and the output waveform is not an exact reproduction of the input waveform (poor fidelity).

The magnetic amplifier is important, however, to many phases of naval engineering because it provides a rugged, trouble-free device that has many applications aboard ship and in aircraft. These applications include throttle controls on the main engines of ships; speed, frequency, voltage, current, and temperature controls on auxiliary equipment; and fire control, servomechanisms, and stabilizers for guns, radar, and sonar equipment.

As stated earlier, the magnetic amplifier does not amplify magnetism, but uses electromagnetism to amplify a signal. It is a power amplifier with a very limited frequency response. Technically, it falls into the classification of an audio amplifier; but, since the frequency response is normally limited to 100 hertz and below, the magnetic amplifier is more correctly called a low-frequency amplifier.

The basic principle of a magnetic amplifier is very simple. (Remember, all amplifiers are current-control devices.) A magnetic amplifier uses a changing inductance to control the power delivered to a

BASIC OPERATION OF A MAGNETIC AMPLIFIER

Figure 3-29 shows a simple circuit with a variable inductor in series with a resistor (representing a load). The voltage source is 100 volts at 60 hertz.

load.
100V 60Hz
E

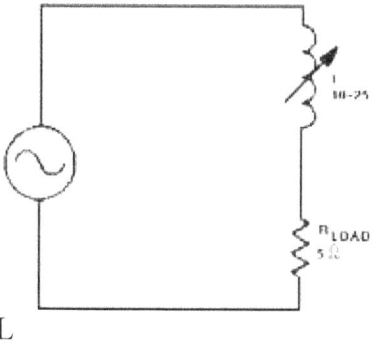

L
10-25bH

Figure 3-29.—Variable inductor in series with a load.

What happens when the inductance decreases ? The end result is that the power in the load (true power) increases . Why? Look at the following formulas and see how each is affected by a decrease in inductance.

X L =27rfL (inductive reactance
in the circuit)
Z=iJX L +R (impedance in the circuit) E
I = — (cunent in the circuit)
Z

True power = I R (true power or power in the load)

(True power is covered in NEETS, Module 2 — Introduction to Alternating Current and Transformers.)

As inductance (L) decreases, X L decreases. As X L decreases, Z decreases. As Z decreases, I increases. Finally, as I increases, true power increases.

This general conclusion can be confirmed by using some actual values of inductance in the formulas along with other values from figure 3-29.

If the value of inductance is 23 millihenries, the formulas yield the following values:

X L = 2a£L
X L = (2)(3.14)(60Hz)(23rnH) X L =8.67Q (rounded off)
Z=^X L 2 +R 2
Z = J(8.67Q) 2 +(5Q) 2
Z = V100.1689Q
Z =10Q (rounded off)
Z
j = ioov
10Q I =10 A
True Power = I 2 R
True Power = (10A) 2 +(5Q) True Power = 500 watts

Now, if the value of inductance is decreased to 11.7 millihenries, the formulas yield the following values:

X L = 2j£L
X L = (2)(3.14)(60Hz)(117rnH) X L = 441Q (rounded off)
Z = ^X L 2 +R 2
Z = -J(4.41&) 2 +(5Q) Z
Z = 6,670. (rounded off)

$I = \dfrac{E}{Z}$

j = 100V 6.67Q

I =15A (rounded off)

True Power = I 2 R

True Power = (ISA) 2 + (5Q) True Power = 1125 watts

So a decrease in inductance of 11.3 millihenries (23 mH—11.7 mH) causes an increase in power to the load (true power) of 625 watts (1125 W—500 W). If it took 1 watt of power to change the inductance by 11.3 millihenries (by some electrical or mechanical means), figure 3-29 would represent a power amplifier with a gain of 625.

Q-39. What is the frequency classification of a magnetic amplifier? Q-40. What is the basic principle of a magnetic amplifier?

Q-41. If inductance increases in a series LR circuit, what happens to true power?

METHODS OF CHANGING INDUCTANCE

Since changing the inductance of a coil enables the control of power to a load, what methods are available to change the inductance? Before answering that question, you should recall a few things about

magnetism and inductors from NEETS, Module I — Introduction to Matter, Energy, and Direct Current, chapter 1— Matter, Energy, and Electricity; and Module 2 — Introduction to Alternating Current and Transformers, chapter 2— Inductance.

Permeability was defined as the measure of the ability of a material to act as a path for additional magnetic lines of force. Soft iron was presented as having high permeability compared with air. In fact, the permeability of unmagnetized iron is 5000 while air has a permeability of 1. A nonmagnetized piece of iron has high permeability because the tiny molecular magnets (Weber's Theory) or the directions of electron spin (Domain Theory) are able to be aligned by a magnetic field. As they align, they act as a path for the magnetic lines of force.

Earlier NEETS modules state that the inductance of a coil increases directly as the permeability of the core material increases. If a coil is wound around an iron core, the permeability of the core is 5000. Now, if the iron is pulled part way out of the coil of wire, the core is part iron and part air. The permeability of the core decreases. As the permeability of the core decreases, the inductance of the coil decreases. This increases the power delivered to the load (true power). This relationship is shown in figure 3-30.

The system shown in figure 3-30 is not too practical. Even if a motor were used in place of the hand that is shown, the resulting amplifier would be large, expensive, and not easily controlled. If the permeability of a core could be changed by electrical means rather than mechanical, a more practical system would result.

DECREASED PERMEABILITY = DECREASED INDUCTANCE

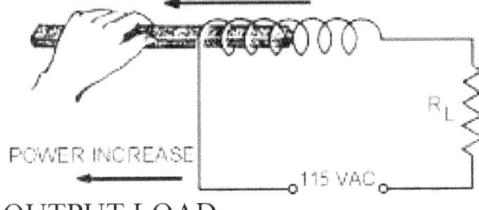

OUTPUT LOAD
SUPPLY

Figure 3-30.—Varying coil inductance with a movable coil.

High permeability depends on there being many molecular magnets (or electron spin directions) that can be aligned to provide a path for magnetic lines of force. If almost all of these available paths are already being used, the material is magnetized and there are no more paths for additional lines of force. The "flux density" (number of lines of force passing through a given area) is as high as it can be. This means that the permeability of the material has decreased. When this condition is reached, the core is said to be SATURATED because it is saturated (filled) with all the magnetic lines of force it can pass. At this point, the core has almost the same value of permeability as air (1) instead of the much higher value of permeability (5000) that it had when it was unmagnetized.

Of course, the permeability does not suddenly change from 5000 to 1. The permeability changes as the magnetizing force changes until saturation is reached. At saturation, permeability remains very low no matter how much the magnetizing force increases. If you were to draw a graph of the flux density compared to the magnetizing force, you would have something similar to the graph shown in figure 3-31. Figure 3-31 also includes a curve representing the value of permeability as the magnetizing force increases. Point "s" in figure 3-31 is the point of saturation. The flux density does not increase above point "s," and the permeability is at a steady, low value.

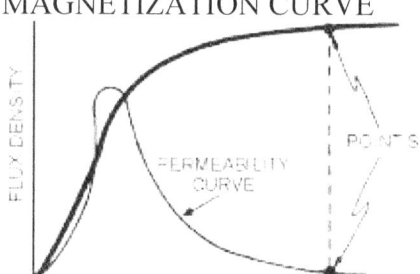

Figure 3-31.—Magnetization and permeability curves.

You have now seen how a change in the magnetizing force causes a change in permeability. The next question is, how do you change the magnetizing force? Magnetizing force is a function of AMPERE-TURNS. (An ampere-turn is the magnetomotive force developed by 1 ampere of current flowing in a coil of one turn.) If you increase the ampere-turns of a coil, the magnetizing force increases. Since it is not practical to increase the number of turns, the easiest way to accomplish this is to increase the current through the coil.

If you increase the current through a coil, you increase the ampere-turns. By increasing the ampere-turns you increase the magnetizing force. At some point, this causes a decrease in the permeability of the core. With the permeability of the core decreased, the inductance of the coil decreases. As said before, a decrease in the inductance causes an increase in power through the load. A device that uses this arrangement is called a SATURABLE-CORE REACTOR or SATURABLE REACTOR.

SATURABLE-CORE REACTOR

A saturable-core reactor is a magnetic-core reactor (coil) whose reactance is controlled by changing the permeability of the core. The permeability of the core is changed by varying a unidirectional flux (flux in one direction) through the core.

Figure 3-32 shows a saturable-core reactor that is used to control the intensity of a lamp. Notice that two coils are wound around a single core. The coil on the left is connected to a rheostat and a battery. This coil is called the control coil because it is part of the control circuit. The coil on the right is connected to a lamp (the load) and an a.c. source. This coil is called the load coil because it is part of the load circuit.

As the wiper (the movable connection) of the rheostat is moved toward the right, there is less resistance in the control circuit. With less resistance, the control-circuit current increases. This causes the amount of magnetism in the core to increase and the inductance of the coil in the load circuit to decrease (because the core is common to both coils). With less inductance in the load circuit, load current increases and the lamp gets brighter.

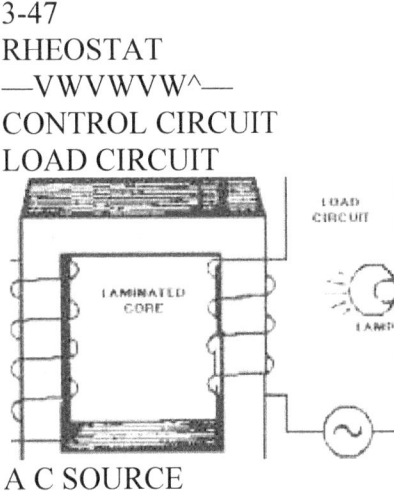

Figure 3-32.—A simple saturable-core reactor circuit.

The schematic diagram of this circuit is shown in figure 3-33. L1 is the schematic symbol for a saturable-core reactor. The control winding is shown with five loops, and the load winding is shown with three loops. The double bar between the inductors stands for an iron core, and the symbol that cuts across the two windings is a saturable-core symbol indicating that the two windings share a saturable core.

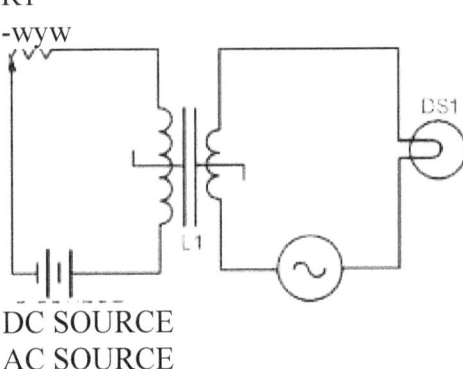

Figure 3-33.—Schematic diagram of a simple saturable-core reactor.

Now that you have seen the basic operation of a saturable-core reactor, there is one other idea to discuss before moving on to the circuitry of a magnetic amplifier. There is a point upon the magnetization curve where the saturable-core reactor should be operated. The ideal operating point is the place in which a small increase in control current will cause a large increase in output power and a small decrease in control current will cause a large decrease in output power. This point is on the flattest portion of the permeability curve (after its peak).

Figure 3-34 shows the magnetization and permeability curves for a saturable-core reactor with the ideal operating point (point "O") indicated. Notice point "O" on the magnetization curve. The portion of the magnetization curve where point "O" is located is called the KNEE OF THE CURVE. The knee of the curve is the point of maximum curvature. It is called the "knee" because it looks like the knee of a leg that is bent. Saturable-core reactors and magnetic amplifiers should be operated on the knee of the magnetization curve.

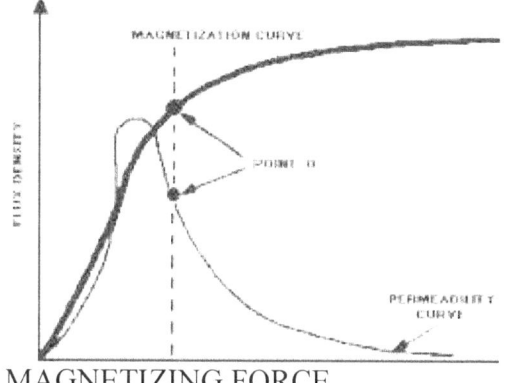

MAGNETIZING FORCE

Figure 3-34.—Magnetization and permeability curves with operating point.

When the saturable-core reactor is set at the knee of the magnetization curve, any small increase in control current will cause a large increase in load current. Any small decrease in control current will cause a large decrease in load current. That is why point "O" is the ideal operating point-because small changes in control current will cause large changes in load current. In other words, the saturable-core reactor can amplify the control current. However, a saturable-core reactor is NOT a magnetic amplifier. You will find out a little later how a magnetic amplifier differs from a saturable-core reactor. First you should know a few more things about the saturable-core reactor.

If a d.c. voltage is applied to the control winding of a saturable-core reactor and an a.c. voltage is applied to the load windings, the a.c. flux will aid the d.c. flux on one half cycle and oppose the d.c. flux on the other half cycle. This is shown in figure 3-35. Load flux is indicated by the dashed-line arrows, and control flux is indicated by the solid-line arrows. View (A) shows the load and control flux adding during one half cycle of the a.c. View (B) of the figure shows the load and control flux opposing during the other half cycle of the a.c.

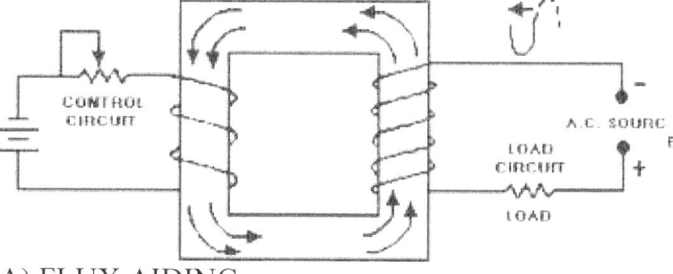

(A) FLUX AIDING

Figure 3-35A.—Flux paths in a saturable-core reactor. FLUX AIDING

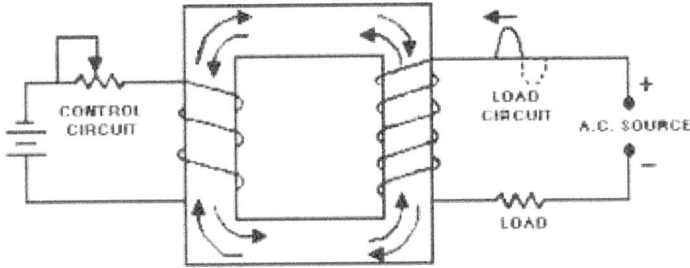

(B) FLUX OPPOSING

Figure 3-35B.—Flux paths in a saturable-core reactor. FLUX OPPOSING

This situation causes the operating point of the saturable-core reactor to shift with the applied a.c. However, the situation would be better if the load flux was not an influence on the control flux. Figure 3-36 shows a circuit in which this is accomplished.

During the first half cycle, the load circuit flux (dashed-line arrows) cancels in the center leg of the core. This is shown in figure 3-36, view (A). As a result, there is no effect upon the flux from the control circuit. During the second half cycle, the polarity of the a.c. (and therefore the polarity of the flux) reverses as shown in view (B). The result is the same as it was during the first half cycle. There is no effect upon the control circuit flux.

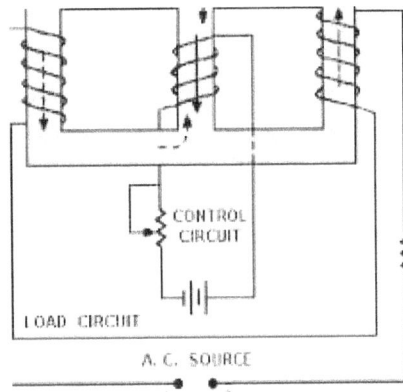

(A) FIRST HALF CYCLE

Figure 3-36A.—Three-legged, saturable-core reactor. FIRST HALF CYCLE

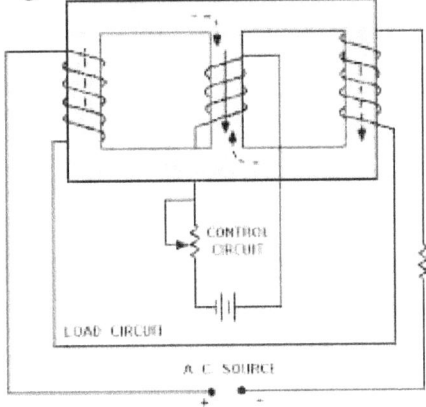

LOAD

(B) SECOND HALF CYCLE

Figure 3-36B.—Three-legged, saturable-core reactor. SECOND HALF CYCLE

Another approach to solving the problem of load flux affecting control flux is shown in figure 3-37. Figure 3-37 shows a toroidal saturable-core reactor. The shape of these cores is a toroid (donut shape). The windings are wound around the cores so that the load flux aids the control flux in one core and opposes the control flux in the other core.

During the first half cycle, the flux aids in the left core and opposes in the right core, as shown in figure 3-37, view (A). During the second half cycle, the flux opposes in the left core and aids in the right core, as shown in view (B). Regardless of the amount of load flux or polarity of the load voltage, there is no net effect of load flux on control flux.

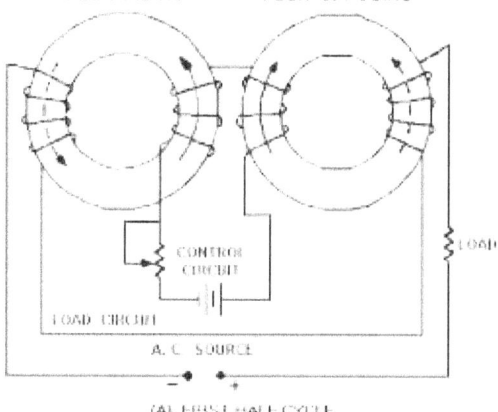

Figure 3-37A.—Toroidal saturable-core reactor. FIRST HALF CYCLE
FLUX OPPOSING FLUX AIDING

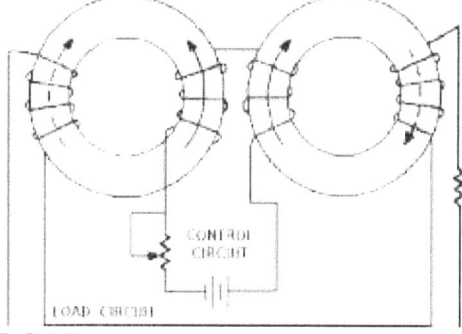

LOAD
A. C. SOURCE
(B) SECOND HALF CYCLE
Figure 3-37B.—Toroidal saturable-core reactor. SECOND HALF CYCLE

Figures 3-36 and 3-37 both represent practical, workable saturable-core reactors. Circuits similar to these are actually used to control lighting in auditoriums or electric industrial furnaces. These circuits are sometimes referred to as magnetic amplifiers, but that is NOT technically correct. A magnetic amplifier differs from a saturable-core reactor in one important aspect: A magnetic amplifier has a rectifier in addition to a saturable-core reactor.

Q-42. If the permeability of the core of a coil increases, what happens to (a) inductance and (b) true power in the circuit?

Q-43. What happens to the permeability of an iron core as the current increases from the operating point to a large value?

Q-44. If two coils are wound on a single iron core, what will a change in current in one coil cause in the other coil?

Q-45. What symbol in figure 3-33 indicates a saturable core connecting two windings?

SIMPLIFIED MAGNETIC AMPLIFIER CIRCUITRY

If the saturable-core reactor works, why do we need to add a rectifier to produce a magnetic amplifier? To answer this question, recall that in NEETS, Module 2 — Introduction to Alternating Current and Transformers, you were told about hysteresis loss. Hysteresis loss occurs because the a.c. applied to a coil causes the tiny molecular magnets (or electron-spin directions) to realign as the polarity of the a.c. changes. This realignment uses up power. The power that is used for realignment is a loss as far as the rest of the circuit is concerned. Because of this hysteresis loss in the saturable-core reactor, the power gain is relatively low. A rectifier added to the load circuit will eliminate the hysteresis loss and increase the gain. This is because the rectifier allows current to flow in only one direction through the load coils.

A simple half-wave magnetic amplifier is shown in figure 3-38. This is a half-wave magnetic amplifier because it uses a half-wave rectifier. During the first half cycle of the load voltage, the diode conducts and the load windings develop load flux as shown in view (A) by the dashed-line arrows. The

load flux from the two load coils cancels and has no effect on the control flux. During the second half cycle, the diode does not conduct and the load coils develop no flux, as shown in view (B). The load flux never has to reverse direction as it did in the saturable-core reactor, so the hysteresis loss is eliminated.

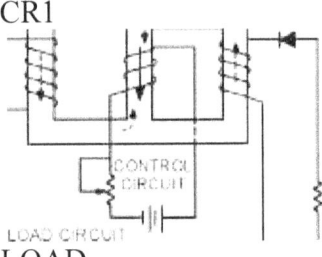

CR1

LOAD
AC SOURCE o o +
(A) FIRST HALF CYCLE

Figure 3-38A.—Simple half-wave magnetic amplifier. FIRST HALF CYCLE

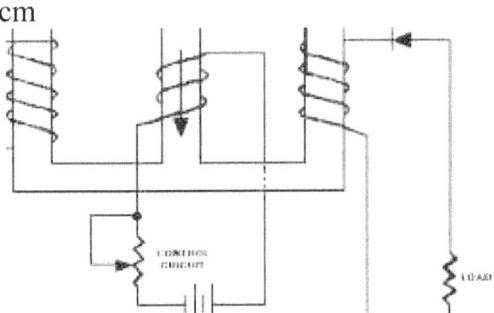

cm

LOAD CIRCUIT
(B) SECOND HALF CYCLE

Figure 3-38B.—Simple half-wave magnetic amplifier. SECOND HALF CYCLE

The circuit shown in figure 3-38 is only able to use half of the load voltage (and therefore half the possible load power) since the diode blocks current during half the load-voltage cycle. A full-wave rectifier used in place of CR1 would allow current flow during the entire cycle of load voltage while still preventing hysteresis loss.

Figure 3-39 shows a simple full-wave magnetic amplifier. The bridge circuit of CR1, CR2, CR3, CR4 allows current to flow in the load circuit during the entire load voltage cycle, but the load current is always in the same direction. This current flow in one direction prevents hysteresis loss.

View (A) shows that during the first half cycle of load voltage, current flows through CR1, the load coils, and CR3. View (B) shows that during the second half cycle, load current flows through CR2, the load coils, and CR4.

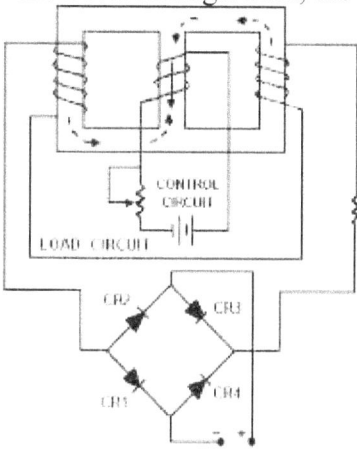

LOAD
A. C. INPUT
(A) FIRST HALF CYCLE

Figure 3-39A.—Simple full-wave magnetic amplifier. FIRST HALF CYCLE

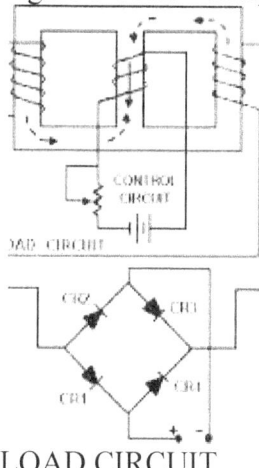

LOAD CIRCUIT
LOAD
A. C. INPUT (B) SECOND HALF CYCLE

Figure 3-39B.—Simple full-wave magnetic amplifier. SECOND HALF CYCLE

Up to this point, the control circuit of the magnetic amplifier has been shown with d.c. applied to it. Magnetic-amplifier control circuits should accept a.c. input signals as well as d.c. input signals. As shown

earlier in figure 3-34, a saturable-core reactor has an ideal operating point. Some d.c. must always be applied to bring the saturable core to that operating point. This d.c. is called BIAS, the most effective way to apply bias to the saturable core and also allow a.c. input signals to control the magnetic amplifier is to use a bias winding. A full-wave magnetic amplifier with a bias winding is shown in figure 3-40.

CONTROL * INPUT 9

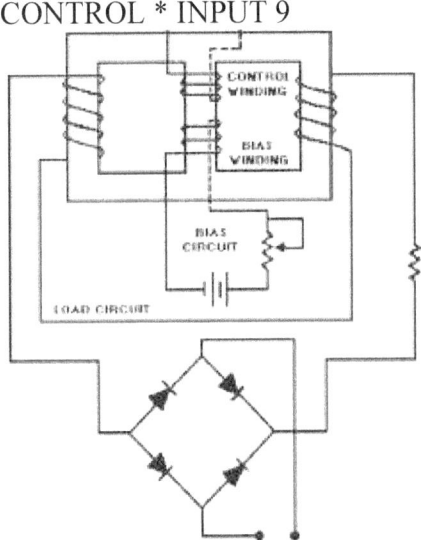

A.C. INPUT

Figure 3-40.—Full-wave magnetic amplifier with bias winding.

In the circuit shown in figure 3-40, the bias circuit is adjusted to set the saturable-core reactor at the ideal operating point. Input signals, represented by the a.c. source symbol, are applied to the control input. The true power of the load circuit is controlled by the control input signal (a.c.)

The block diagram symbol for a magnetic amplifier is shown in figure 3-41. The triangle is the general symbol for an amplifier. The saturable-core reactor symbol in the center of the triangle identifies the amplifier as a magnetic amplifier. Notice the input and output signals shown. The input signal is a small-amplitude, low-power a.c. signal. The output signal is a pulsating d.c. with an amplitude that varies. This variation is controlled by the input signal and represents a power gain of 1000.

INPUT SIGNAL
1V 500mW

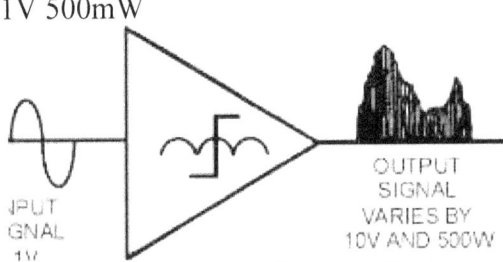

Figure 3-41.—Magnetic amplifier input and output signals.

Some magnetic amplifiers are designed so a.c. goes through the load rather than pulsating d.c. This is done by placing the load in a different circuit position with respect to the rectifier. The principle of the magnetic amplifier remains the same: Control current still controls load current.

Magnetic amplifiers provide a way of accurately controlling large amounts of power. They are used in servosystems (which are covered later in this training series), temperature or pressure indicators, and power supplies.

This chapter has presented only the basic operating theory of saturable-core reactors and magnetic amplifiers. For your convenience, simple schematic diagrams have been used to illustrate this material. When magnetic amplifiers and saturable-core reactors are used in actual equipment, the schematics may be more complex than those you have seen here. Also, you may find coils used in addition to those presented in this chapter. The technical manual for the equipment in question should contain the information you need to supplement what you have read in this chapter.

Q-46. At what portion of the magnetization curve should a magnetic amplifier be operated? Q-47. How is the effect of load flux on control flux eliminated in a saturable-core reactor? Q-48. What is the purpose of the rectifier in a magnetic amplifier?

Q-49. What is used to bias a magnetic amplifier so that the control winding remains free to accept control (input) signals?

Q-50. List two common usages of magnetic amplifiers.

This chapter has presented information on differential amplifiers, operational amplifiers, and magnetic amplifiers. The information that follows summarizes the important points of this chapter.

A DIFFERENCE AMPLIFIER is any amplifier with an output signal dependent upon the difference between the input signals. A two-input, single-output difference amplifier can be made by combining the common-emitter and common-base configurations in a single transistor.

SUMMARY

+ vcc

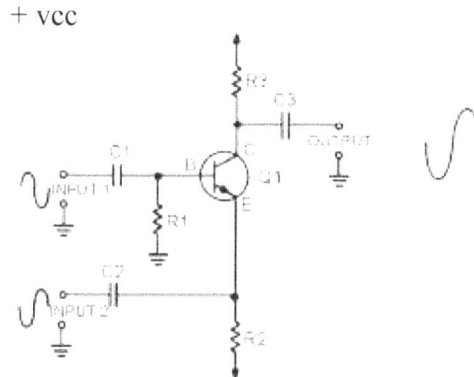

-Vee

A difference amplifier can have input signals that are IN PHASE with each other, view (A), 180 DEGREES OUT OF PHASE with each other, view (B), or OUT OF PHASE BY SOMETHING OTHER THAN 180 DEGREES with each other, view (C).

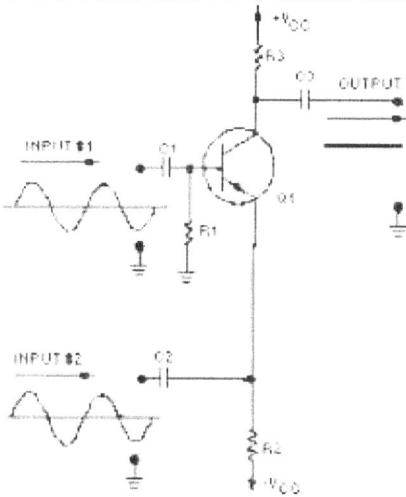

(A) DIFFERENTIAL AMPLIFIER IN PHASE

± :;ri
INPUT *1 ? 1 ^ H
1

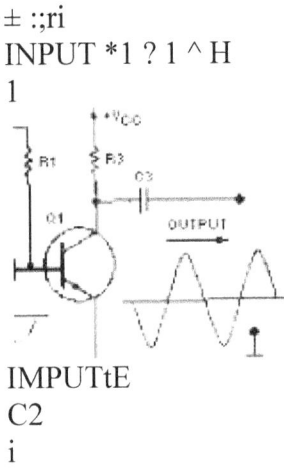

IMPUTtE
C2
i
(B) DIFFERENTIAL AMPLIFIER 180° OUT OF PHASE

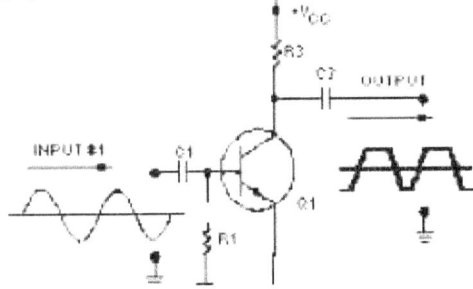

INFUT*£
:r2
(Q DIFFERENTIAL AMPLIFIER OUT OF PHASE OTHER THAN 180

A DIFFERENTIAL AMPLIFIER has two possible inputs and two possible outputs. The combined output signal is dependent upon the difference between the input signals.

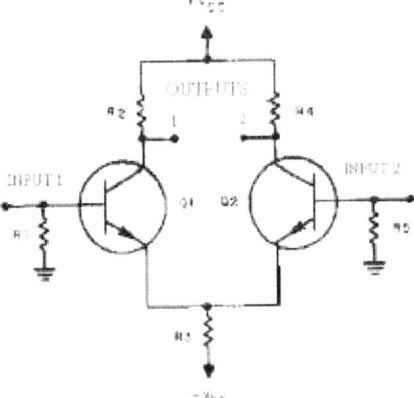

A differential amplifier can be configured with a SINGLE INPUT and a SINGLE OUTPUT

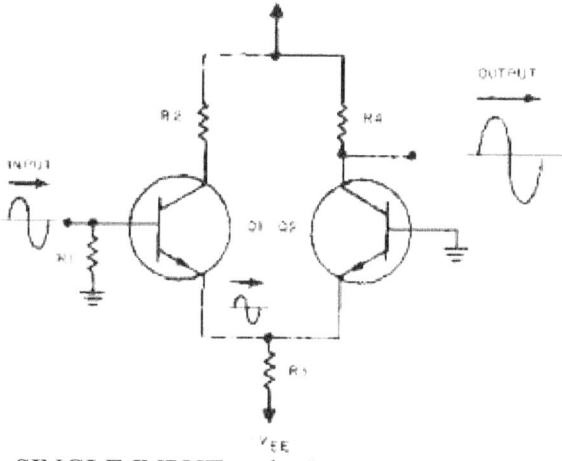

a SINGLE INPUT and a DIFFERENTIAL OUTPUT
COMBINED DIFFERENTIAL OUTPUT

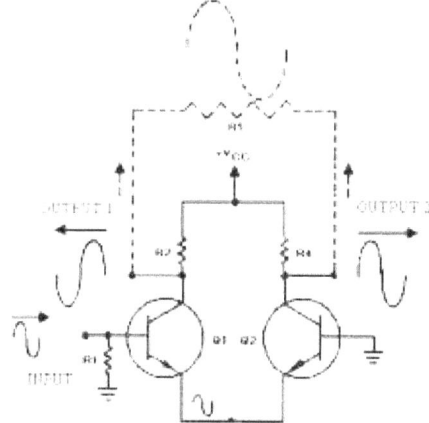

R3
or a DIFFERENTIAL INPUT and a DIFFERENTIAL OUTPUT.
COMBINED OUTPUT

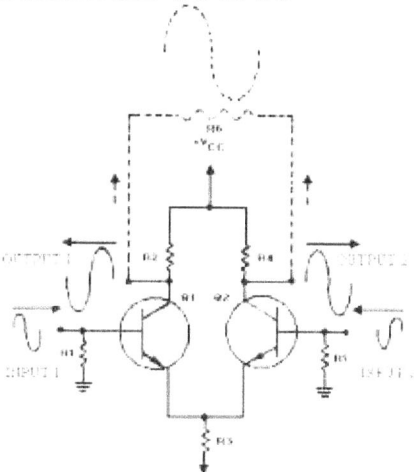

V EE

An OPERATIONAL AMPLIFIER is an amplifier which has very high gain, very high input impedance, and very low output impedance. An OP AMP is made from three stages: (1) a differential amplifier, (2) a high-gain voltage amplifier, and (3) an output amplifier.

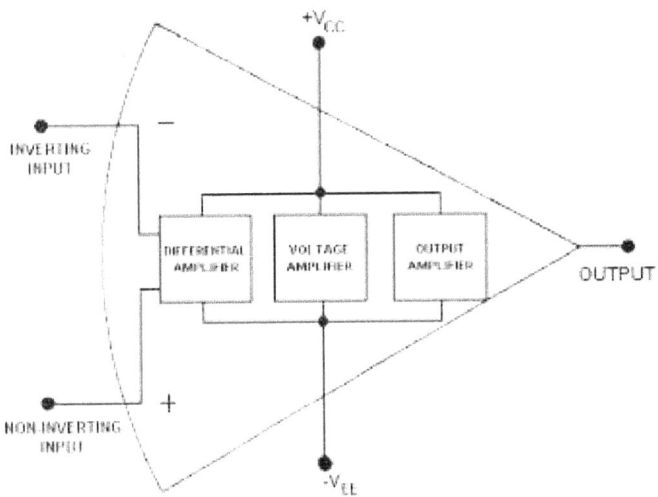

Operational amplifiers are usually used in a CLOSED-LOOP OPERATION. This means that degenerative feedback is used to lower the gain and increase the stability of the operational amplifier.

An operational amplifier circuit can be connected with an INVERTING CONFIGURATION

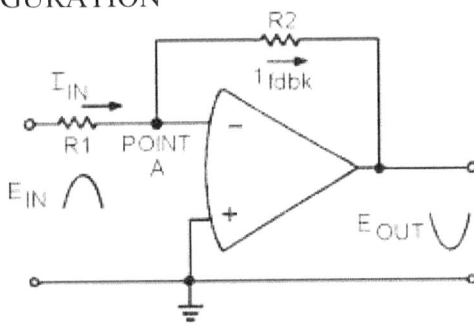

or a NONINVERTING CONFIGURATION.

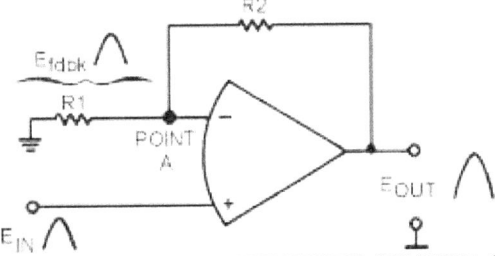

The GAIN-BANDWIDTH PRODUCT for an operational amplifier is computed by multiplying the gain by the bandwidth (in hertz). For any given operational amplifier, the gain-bandwidth product will remain the same regardless of the amount of feedback used.

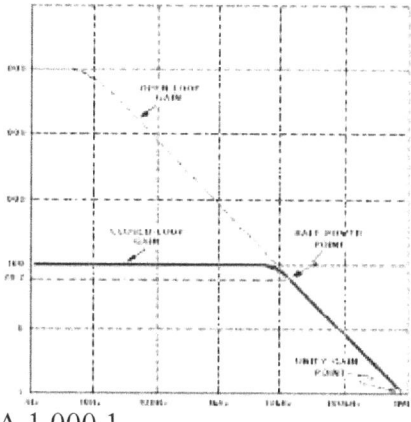

A 1.000 1-
100Hz 1kHz 10kHz
FREQUENCY
100kHz 1MHz
1Hz 10Hz 100Hz 1kHz 10kHz 100kHz 1MHz
FREQUENCY

A SUMMING AMPLIFIER is an application of an operational amplifier in which the output signal is determined by the sum of the input signals multiplied by the gain of the amplifier:

E out = gain (E1 + E2 ...)

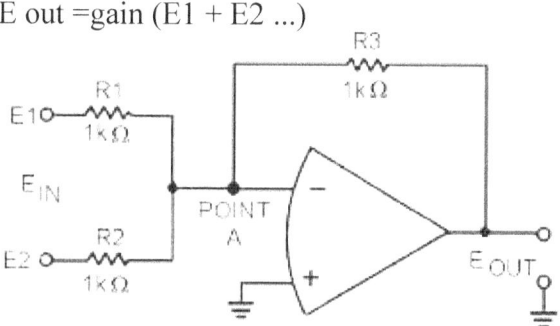

A SCALING AMPLIFIER is a special type of summing amplifier with the output signal determined by multiplying each input signal by a different factor (determined by the ratio of the input-signal resistor and feedback resistor) and then adding these products:

J out
R.
in1
R
in2
+ (^dbk xE3)...]

A DIFFERENCE AMPLIFIER is an application of an operational amplifier in which the output signal is determined by the difference between the input signals multiplied by the gain of the amplifier:

E out = g ain (E2-E1)

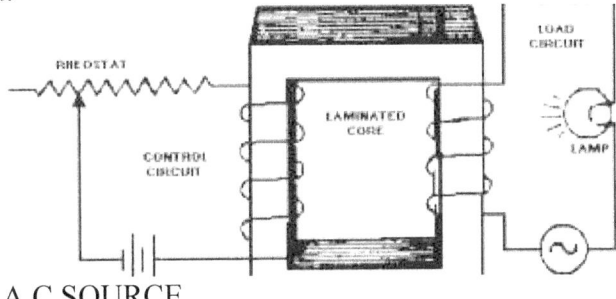

A SATURABLE-CORE REACTOR works upon the principle that increasing the current through a coil decreases the permeability of the core; the decreased permeability decreases the inductance of the coil which causes an increase in current (power) through the load.

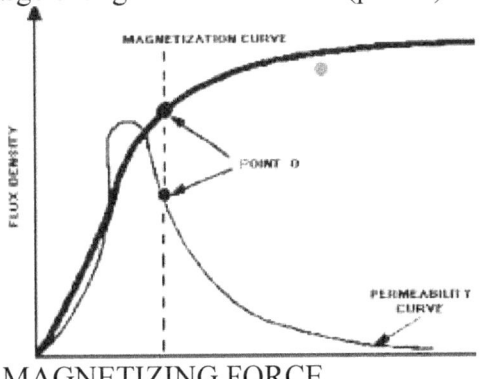

A C SOURCE

The IDEAL OPERATING POINT of a saturable-core reactor is on the KNEE OF THE MAGNETIZATION CURVE. At this point, small changes in control current will cause large changes in load current (power).

MAGNETIZING FORCE

THREE-LEGGED and TOROIDAL saturable-core reactors solve the problem of load flux aiding and opposing control flux during alternate half cycles of the a.c. load current.

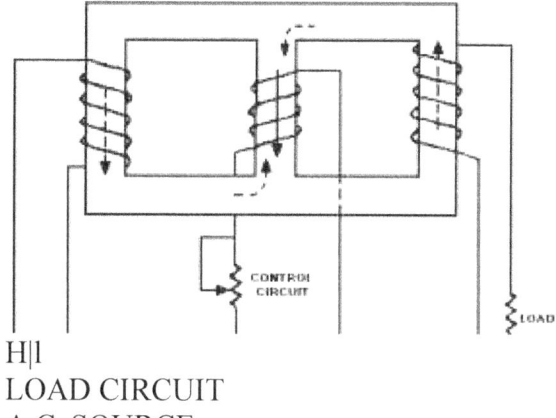

H|l
LOAD CIRCUIT
A.C. SOURCE —••—

MAGNETIC AMPLIFIERS use the principle of electromagnetism to amplify signals. They are power amplifiers with a frequency response normally limited to 100 hertz or below. Magnetic amplifiers use a saturable-core reactor. A magnetic amplifier uses a RECTIFIER to solve the problem of HYSTERESIS LOSS in a saturable-core reactor.

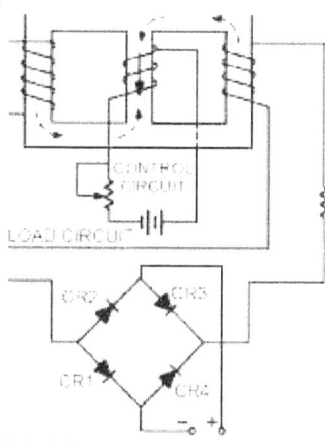

LOAD
AC CIRCUIT

A BIAS WINDING allows a d.c. bias voltage to be applied to the saturable-core reactor while a.c. control signals are applied to a separate control winding. In this way a magnetic amplifier can be set to the proper operating point.
CONTROL • INPUT f

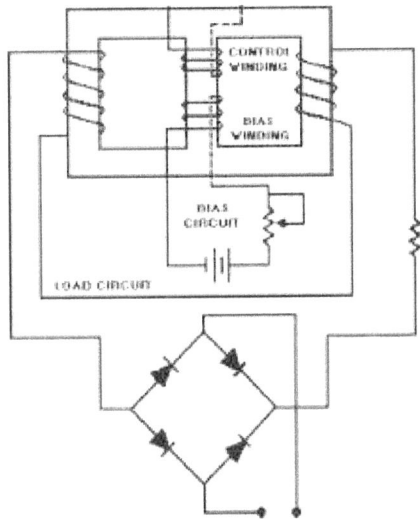

A.C. INPUT

ANSWERS TO QUESTIONS Q1. THROUGH Q50.

A-1. Two inputs, two outputs.

A-2. Common emitter (CE) and common base (CB). A-3. No output (the signals will "cancel out").

A-4. Equal in shape and frequency to each input signal and larger in amplitude by two times than either input signal.

A-5. Equal in shape and frequency to the input signal; larger in amplitude than the input signal; half as large in amplitude as when two input signals were used that were 180 degrees out of phase.

A-6. A different shape than the input signals but larger in amplitude.

A-7. 100 millivolts.

A-8. Each output will be a sine wave with a peak-to-peak amplitude of 100 millivolts. The output signals will be 180 degrees out of phase with each other.

A-9. 200 millivolts.

A-10. 0 volts (the input signals will "cancel out").

A-11. Each output signal will be 100 millivolts.

A-12.

a. 180 degrees out of phase with each other.

b. Output signal number one will be in phase with input signal number two; output signal number two will be in phase with input signal number one.

A-13. 200 millivolts.

A-14.

a. 100 millivolts.

b. No.

A-15. Very high gain, very high input impedance, very low output impedance. A-16. An integrated circuit (chip).

A-17.

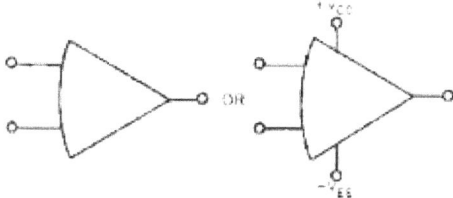

A-18.
a. Differential amplifier.
b. Voltage amplifier.
c. Output amplifier.

A-19. The use of degenerative (negative) feed-back. A-20. Both the input signal and the feedback signal. A-21.
a. Inverting.
b. Inverting. A-22. 0 volts. A-23. Virtual. A-24. -50 millivolts.

A-25. 50 kilohertz (Gain = 10; Gain-Bandwidth Product = 500,000; 50ftOOO(H,) 10

A-26. 60 millivolts.

A-27. 1 megahertz.
Open-loop Gain-Bandwidth Product = Closed-loop Gain-Bandwidth Prod.
Open-loop Gain-Bandwidth Product = 200,000 x30 (Hz)
Open-loop Gain Bandwidth Product = 600,000
Closed-loop Gain Bandwidth Product = 6 x Bandwidth
6,000,000 = 6 xBandwidth
1,000,000 (Hz) = Bandwidth

A-28. The adder simply adds the input signals together while the summing amplifier multiplies the sum of the input signals by the gain of circuit.

A-29. Yes, a summing amplifier can have as many inputs as desired.

A-30. A summing amplifier that applies a factor to each input signal beforeadding the results.

A-31. A scaling amplifier.

A-32.
E 0Ut =-72V
Solution:
, ,,, 30kQ, . ,,, 30kfi. E out = - +2V x + +6Vx
out 5kQ 3kQ
E 01]t =-(+2Vx6) + (+6VxlO)

A-33. 0 volts. (The two inputs to the operational amplifier are both at 0 volts.)

A-34. The difference amplifier multiplies the difference between the two inputs by the gain of the circuit while the subtractor merely subtracts one input signal from the other.

A-35. No.

A-36. A difference amplifier.

A-37.
E
= +60V
Solution:

E 01jt =(+6V) +
20kQ 2kQ
E out =(+6V)xlO

A-38. 0 volts. (The two inputs to the operational amplifier are both at the same potential.)

A-39. An audio (or low) frequency power amplifier.

A-40. A change in inductance in a series LR circuit causes a change in true power.

A-41. It decreases.

A-42. (a) Inductance increases; (b) true power decreases.

A-43. Permeability decreases.

A-44. A change in inductance.

A-46. The knee of the curve.

A-47. Use two load windings whose flux effects cancel in the core of the reactor or use two load windings on two toroidal cores so that load flux always aids control flux in one core and opposes control flux in the other core.

A-48. The rectifier eliminates hysteresis loss.

A-49. A bias winding and associated circuitry.

A-50. Servosystems, temperature recorders, or power supplies.

A-45.

APPENDIX I

GLOSSARY

AMPERE-TURN—The magnetomotive force developed by one ampere of current flowing through a coil of one turn.

AMPLIFICATION—The process of enlarging a signal in amplitude (as of voltage or current).

AMPLIFIER—A device that enables an input signal to control an output signal. The output signal will have some (or all) of the characteristics of the input signal but will generally be larger than the input signal in terms of voltage, current, or power.

AMPLITUDE—The size of a signal. Generally used to describe voltage, current, or power.

AUDIO AMPLIFIER—An amplifier designed to amplify frequencies between 15 hertz (15 Hz) and 20 kilohertz (20 kHz).

BANDWIDTH—The difference between the highest usable frequency of a device (upper frequency limit) and the lowest usable frequency of the device (lower frequency limit)—measured at the half-power points.

BYPASS CAPACITOR—A capacitor used to transfer unwanted signals out of a circuit; e.g., coupling an unwanted signal to ground. Also called a DECOUPLING CAPACITOR.

COMBINATION PEAKING—A technique in which a combination of peaking coils in series and parallel (shunt) with the output signal path is used to improve high-frequency response.

COMPENSATION—The process of overcoming the problems associated with frequencies in an amplifier.

COUPLING—The process of transferring energy from one point in a circuit to another point or from one circuit to another.

COUPLING CAPACITOR—A capacitor used to couple signals.

DECOUPLING CAPACITOR—A capacitor used to transfer unwanted signals out of a circuit; e.g., coupling an unwanted signal to ground. Also called a BYPASS CAPACITOR.

DEGENERATIVE FEEDBACK—Feedback in which the feedback signal is out of phase with the input signal, also called NEGATIVE FEEDBACK.

DIFFERENTIAL AMPLIFIER—An amplifier with an output which is determined by the difference between two input signals and which can provide up to two output signals.

DISTORTION—Any unwanted change between an input signal and an output signal.

DRIVER—An electronic circuit that supplies the input to another circuit.

FEEDBACK—The process of sending part of an output signal of an amplifier back to the input of the amplifier.

AI-1

FIDELITY—The quality of reproducing an output signal exactly like the input signal except for

amplitude (and somtimes phase); i.e., output and input signals exactly alike in terms of frequency and shape.

FREQUENCY-DETERMINING NETWORK—A circuit that provides the desired response (maximum or minimum impedance) at a specific frequency.

FREQUENCY-RESPONSE CURVE—A curve showing the output of an amplifier (or any other

device) in terms of voltage or current plotted against frequency with a fixed-amplitude input signal.

GAIN-BANDWIDTH PRODUCT—The number that results when the gain of a circuit is multiplied by the bandwidth of that circuit. For an operational amplifier, the gain-bandwidth product for one configuration will always equal the gain-bandwidth product for any other configuration of the same amplifier.

HALF-POWER POINTS—The points on a frequency-response curve at which the output power is one-half of the maximum power out.

HIGH-FREQUENCY COMPENSATION—See peaking coil.

KNEE OF THE CURVE—The point of maximum curvature. (Shaped like the knee of a leg that is bent.)

MAGNETIC AMPLIFIER (MAG AMP)—An amplifier that uses electromagnetic effects to provide amplification of a signal. The magnetic amplifier uses a changing inductance to control the power delivered to a load.

NEGATIVE FEEDBACK—Feedback in which the feedback signal is out of phase with the input signal. Also called DEGENERATIVE FEEDBACK.

NEUTRALIZATION—The process of counteracting or "neutralizing" the effects of interelectrode capacitance.

OPERATIONAL AMPLIFIER (OP AMP)—An amplifier designed to perform computing or transfer operations and which has the following characteristics: (1) very high gain, (2) very high input impedance, and (3) very low out put impedance.

PEAKING COIL—An inductor used in an amplifier to provide high-frequency compensation which extends the high-frequency response of the amplifier.

PERMEABILITY—The measure of the ability of a material to act as a path for additional magnetic lines of force.

PHASE SPLITTER—A device that provides two output signals from a single input signal. The two output signals will differ from each other in phase.

POSITIVE FEEDBACK—Feedback in which the feedback signal is in phase with the input signal. Also called REGENERATIVE FEEDBACK.

POWER AMPLIFIER—An amplifier in which the output-signal power is greater than the input-signal power.

PUSH-PULL AMPLIFIER—An amplifier which uses two transistors (or electron tubes) whose output signals are combined to provide a larger gain (usually a power gain) than a single transistor (or electron tube) can provide.

AI-2

REGENERATIVE FEEDBACK—Feedback in which the feedback signal is in phase with the input signal. Also called POSITIVE FEEDBACK

RF (RADIO FREQUENCY) AMPLIFIER—An amplifier designed to amplify signals with frequencies between 10 kilohertz (10 kHz) and 100,000 megahertz (100,000 MHz).

RF (RADIO FREQUENCY) TRANSFORMER—A transformer specially designed for use with rf

(radio frequencies). An rf transformer is wound onto a tube of non- magnetic material and has a core of either powdered iron or air.

SATURATION (MAGNETIC CORE)—The condition in which a magnetic material has reached a maximum flux density and the permeability has decreased to a value of (approximately) 1.

SATURABLE-CORE REACTOR—A coil whose reactance is controlled by changing the permeability of the core.

SERIES PEAKING—A technique used to improve high-frequency response in which a peaking coil is placed in series with the output signal path.

SHUNT PEAKING—A technique used to improve high-frequency response in which a peaking coil is placed in parallel (shunt) with the output signal path.

SIGNAL—A general term used to describe any a.c. or d.c. of interest in a circuit; e.g., input signal.

STAGE—One of a series of circuits within a single device; e.g., first stage of amplification.

SWAMPING RESISTOR—A resistor used to increase or "broaden" the bandwidth of a circuit.

TUNED CIRCUIT—An LC circuit used as frequency-determining network.

VIDEO AMPLIFIER—An amplifier designed to amplify the entire band of frequencies from 10 hertz (10 Hz) to six megahertz (6 MHz). Also called a WIDE-BAND AMPLIFIER.

VIRTUAL GROUND—A point in a circuit which is at ground potential (0 V) but is not connected to ground.

VOLTAGE AMPLIFIER—An amplifier in which the output-signal voltage is greater than the input-signal voltage.

WIDE-BAND AMPLIFIER—An amplifier designed to pass an extremely wide band of frequencies, i.e., video amplifier.

AI-3

MODULE 8 INDEX

Amplifier frequency response, 2-2 to 2-11 bandwidth of an amplifier, 2-3 to 2-5 factors affecting frequency response of an

amplifier, 2-9 to 2-11 reading amplifier frequency-response curve, 2-5 to 2-7 Amplifier input/output impedance and gain,

2-18, 2-19 Amplifiers, 1-1 to 1-29 audio, 1-24 to 1-29

phase splitters, 1-24 to 1-26 push-pull amplifiers, 1-27 to 1-28 single-stage, 1-28, 1-29 classification of, 1-25 to 1-26

frequency response of amplifiers, 1-5 voltage amplifiers and power amplifiers, 1-3, 1-4 introduction, 1-1, 1-2 summary, 1-29 to 1-38 transistor, 1-5 to 1-24

classes of operation, 1-6 to 1-9 coupling, 1-9 to 1-13 feedback, 1-17 to 1-24 impedance considerations for, 1-14 to 1-17

Amplifiers, special, 3-1 to 3-57

differential amplifiers, 3-2 to 3-14

basic differential amplifier circuit, 3-2 differential-input, differential-output,

differential amplifier, 3-13 single-input, differential-output,

differential amplifier, 3-12 single-input, single-output, differential

amplifier, 3-11 two-input, single-output, difference

amplifier, 3-3 to 3-10 typical differential amplifier circuit, 3-10 introduction, 3-1

magnetic amplifiers, 3-42 to 3-57 basic operation, 3-42 to 3-44

Amplifiers, special—Continued

methods of changing inductance, 3-44 to 3-46

saturable-core reactor, 3-46 to 3-52 simplified magnetic amplifier circuitry, 3-52 to 3-57 operational amplifiers, 3-15 to 3-41 applications, 3-29 to 3-40 bandwidth limitations, 3-25 to 3-28 block diagram, 3-16 to 3-18 characteristics, 3-15 closed-loop operation, 3-18 to 3-23 summary, 3-57 to 3-69 Amplifiers, video, 2-12 to 2-16

high-frequency compensation, 2-12 low-frequency compensation, 2-15, 2-16 typical video-amplifier circuit, 2-16 Applications of operational amplifiers, 3-29 to 3-40

Audio amplifiers, 1-24 to 1-26 phase splitters, 1-27, 1-28 push-pull amplifiers, 1-28, 1-29 single-stage, 1-25, 1-26

B

Bandwidth limitations, operational amplifiers,

3-25 to 3-28 Bandwidth of an amplifier, 2-3 to 2-5 Block diagram, operational amplifiers, 3-16 to

3-18

C

Changing inductance, methods of, magnetic

amplifiers, 3-44 to 3-46 Classes of operation, amplifier, 1-6 to 1-9

class A, 1-6, 1-7

class AB, 1-7

class B, 1-7, 1-8

class C, 1-8, 1-9 Classification of amplifiers, 1-3, 1-4

frequency response, 1-5

voltage amplifiers and power amplifiers, 1-3, 1-4

Closed-loop operation, operational amplifiers, 3-18 to 3-23
Combination peaking, 2-14, 2-15
Compensation of rf amplifiers, 2-23 to 2-25
 neutralization, 2-24, 2-25
 transformers, 2-23
Coupling, amplifiers, 1-9 to 1-13
 direct, 1-10, 1-11
 impedance, 1-12, 1-13
 RC, 1-11, 1-12
 transformer, 1-13
Coupling, rf amplifier, 2-21 to 2-23

Differential amplifiers, 3-2 to 3-14
 basic differential amplifier circuit, 3-2
 differential-input, differential-output, differential amplifier, 3-13
 single-input, differential-output, differential amplifier, 3-12
 single-input, single-output, differential amplifier, 3-11
 two-input, single-output, difference amplifier, 3-3 to 3-10
 typical differential amplifier circuit, 3-10
Direct coupling, 1-10, 1-11

Factors affecting frequency response of an amplifier, 2-9 to 2-11
Feedback, amplifier, 1-17 to 1-24
 negative, 1-23, 1-24
 positive, 1-20 to 1-23
Frequency-determining network, rf amplifier, 2-19
Frequency-response curve, reading amplifier, 2-5 to 2-7
Frequency response of amplifiers, 1-5

Glossary, AI-1 to AI-3

H
High-frequency compensation for video, 2-12 to 2-16
 amplifiers, 2-12 to 2-15
 combination peaking, 2-14, 2-15
 series peaking, 2-12, 2-13
 shunt peaking, 2-13, 2-14

Impedance considerations for amplifiers, 1-14 to 1-17
Impedance coupling, 1-12, 1-13

L
Learning objectives, video and rf amplifiers, 2-1
Low-frequency compensation for video amplifiers, 2-15, 2-16

M
Magnetic amplifiers, 3-42 to 3-57
 basic operation, 3-42 to 3-44
 methods of changing inductance, 3-44 to 3-46
 saturable-core reactor, 3-46 to 3-52
 simplified magnetic amplifier circuitry, 3-52 to 3-57

O
Operational amplifiers, 3-15 to 3-41
 applications, 3-29 to 3-40
 difference amplifier (subtractor), 3-36 to 3-40
 summing amplifier (adder), 3-29 to 3-35
 bandwidth limitations, 3-25 to 3-28
 block diagram, 3-16, 3-17
 characteristics, 3-15
 closed-loop operation, 3-18 to 3-28
 inverting configuration, 3-18 to 3-23
 noninverting configuration, 3-23 to 3-28

P

Peaking, high-frequency compensation for video amplifiers, 2-12 to 2-15
  combination, 2-14
  series, 2-13
  shunt, 2-12 Phase splitters, 1-27, 1-28 Power amplifiers and voltage amplifiers, 1-3, 1-4
Push-pull amplifiers, 1-28, 1-29

R

Radio-frequency amplifiers, 2-17 to 2-27
  amplifier input/output impedance and gain, 2-18, 2-19 compensation, 2-23 to 2-25 frequency-determining network, 2-19 rf amplifier coupling, 2-21 to 2-23 typical circuits, 2-25 to 2-27
RC coupling, 1-11, 1-12
Reading amplifier frequency-response curve, 2-9

S

Saturable-core reactor, magnetic amplifiers, 3-46 to 3-52
Series peaking, 2-12, 2-13 Shunt peaking, 2-13, 2-14 Single-stage audio amplifiers, 1-25, 1-26

Transistor amplifiers—Continued
  impedance considerations for amplifiers, 1-14 to 1-17

V

Video and rf amplifiers, 2-1 to 2-27
  amplifier frequency response, 2-5 to 2-7 bandwidth of an amplifier, 2-3 to 2-5 factors affecting frequency response
  of an amplifier, 2-9 to 2-11 reading amplifier frequency-response curve, 2-5 to 2-7 introduction, 2-1 learning objectives, 2-1 radio-frequency amplifiers, 2-17 to 2-27 amplifier input/output impedance and
  gain, 2-18, 2-19 compensation, 2-23 to 2-25 frequency-determining network, 2-19 rf amplifier coupling, 2-21 to 2-23 typical circuits, 2-25 to 2-27 summary, 2-28 to 2-33 video amplifiers, 2-12 to 2-15
    high-frequency compensation, 2-12, 2-13
    low-frequency, 2-15 to 2-16 typical video-amplifier circuit, 2-16 Voltage amplifiers and power amplifiers, 1-3, 1-4
  Transformer coupling, 1-13 Transistor amplifiers, 1-5 to 1-24 classes of operation, 1-6 to 1-9 coupling, 1-9 to 1-13 feedback, 1-17 to 1-24

INDEX-3

Assignment Questions

Information : The text pages that you are to study are provided at the beginning of the assignment questions.

ASSIGNMENT 1

Textbook assignment: Chapter 1, "Amplifiers," pages 1-1 through 1-40. Chapter 2, "Video and RF
Amplifiers," pages 2-1 through 2-34.

1-1. The control of an output signal by an

input signal resulting in the output signal having some (or all) of the characteristics of the input signal is known by which of the following terms?

1. Multiplication
2. Magnification
3. Amplification
4. Addition

1-2. Which of the following statements describes the relationship of input and output signals in a amplifier?

1. The input signal is actually changed into the output signal
2. Both the input and output signal are unchanged; neither is affected by the other
3. The input signal is controlled by the output signal and the output signal remains unchanged
4. The input signal remains unchanged and the output signal is controlled by the input signal

1-3. Why are amplifiers used in electronic devices?

1. To provide signals of usable amplitude
2. To "pick up" broadcast signals
3. To select the proper broadcast signal
4. To change the broadcast signal to an audio signal

1-4. Most amplifiers can be classified in which of the following ways?

1. Function and size
2. Power requirements and size
3. Function and frequency response
4. Frequency response and power requirements

1-5. The speaker system of a record player should be driven by which of the following types of amplifier?

1. An audio power amplifier
2. A video voltage amplifier
3. A direct-current voltage amplifier
4. An alternating-current rf amplifier

1-6. The signal from a radio antenna should be amplified by which of the following types of amplifier?

1. An rf voltage amplifier
2. A video power amplifier
3. A direct-current audio amplifier
4. An alternating-current power amplifier

1-7. The class of operation of an amplifier is determined by which of the following factors?

1. The gain of the stage
2. The efficiency of the amplifier
3. The amount of time (in relation to the input signal) that current flows in the output circuit
4. The amount of current (in relation to the input-signal current) that flows in the output circuit

1-8. Which of the following is NOT a class of operation for an amplifier?

1. A

2. C
3. AB
4. AC

1-9. If the output of a circuit should be a representation of less than one-half of the input signal, what class of operation should be used?
1. A
2. C
3. AB
4. AC

1-10. What class of operation is the most efficient?
1. A
2. C
3. AB
4. AC

1-11. What class of operation has the highest fidelity?
1. A
2. C
3. AB
4. AC

1-12. What is the purpose of an amplifier-coupling network?
1. To "block" d.c.
2. To provide gain between stages
3. To separate one stage from another
4. To transfer energy from one stage to another

1-13. Which of the following is NOT a method of coupling amplifier stages?
1. RC
2. Resistor
3. Impedance
4. Transformer

1-14. What is the most common form of coupling?
1. RC
2. Resistor
3. Impedance
4. Transformer

1-15. Which of the following types of coupling is usually used to couple the output from a power amplifier?
1. RC
2. Resistor
3. Impedance
4. Transformer

1-16. Which of the following types of amplifiers have both high and low frequency response limitations?
1. RC
2. Resistor
3. Impedance

4. Transformer

1-17. Which of the following types of coupling is most effective at high frequencies?

1. RC
2. Resistor
3. Impedance
4. Direct

THIS SPACE LEFT BLANK INTENTIONALLY.

1-18. For maximum power transfer between circuits, what impedance relationship should there be between the two circuits?

1. The output impedance of circuit number one should be higher
2. The input impedance of circuit number one should be higher than the output impedance of circuit .number two
3. The output impedance of circuit number one should be lower than the input impedance of circuit number two
4. The output impedance of circuit number one should be equal to the input impedance of circuit number two

1-19. For maximum current at the input to a circuit, what should the relationship of the input impedance be to the output impedance of the previous stage?

1. Higher than
2. Lower than
3. Equal to
4. The impedance relationship is immaterial

1-20. What is the (a) input impedance and (b) output impedance of a common-base transistor configuration?

1-21. What transistor configuration should b e used to match a high output impedance to a low input impedance?

1. Common collector
2. Common emitter
3. Common gate
4. Common base

1-22. What type of coupling is most useful for impedance matching?

1. RC
2. Resistor
3. Impedance
4. Transformer

1-23. What is feedback?

1. The control of a circuit output signal by the input signal
2. The control of a circuit input signal by the output signal
3. The coupling of a portion of the output signal to the input of the circuit
4. The coupling of a portion of the input signal to the output of the circuit

1-24. Which of the following terms describe the two types of feedback?

1. Positive and negative
2. Degenerative and regenerative
3. Both 1 and 2 above
4. Bypassed and unbypassed

1-25. What type of feedback provide an increased amplitude output signal?
1. Positive
2. Negative
3. Bypassed
4. Unbypassed

1-26. Distortion caused by amplifier saturation can be reduced by using which of the following types of feedback?
1. Positive
2. Negative
3. Regenerative
4. Unbypassed

1-27. What type feedback is provided if the feedback signal is out of phase with the input signal?
1. Unbypassed
2. Bypassed
3. Negative
4. Positive

1-28. What type of feedback is provided by a capacitor connected across the emitter-resistor in a common-emitter transistor amplifier?
1. Bypassed
2. Positive
3. Negative
4. Unbypassed

1-29. What are the (a) inputs and (b) outputs of a phase splitter?
1. (a) Two signals in phase (b) One signal
2. (a) Two signals out of phase (b) One signal
3. (a) One signal
(b) Two signals in phase
4. (a) One signal
(b) Two signals out of phase

1-30. A single-stage, two transistor amplifier that uses a phase splitter input is classified as what type of amplifier?
1. Inverse
2. Push-pull
3. Phase splitter
4. Regenerative

1-31. Which of the following is a common use for a push-pull amplifier?
1. The first stage of a video amplifier
2. The amplifier stage connected directly to an antenna
3. The second stage in a four stage rf amplifier
4. The final stage in an audio amplifier

1-32. What is the advantage of a push-pull amplifier as compared to a single transistor amplifier?
1. Lower cost
2. Fewer parts
3. Higher gain

4. Less power usage

1-33. To provide good fidelity output signals, which of the following classes of operation CANNOT be used by a push-pull amplifier?
1. A
2. B
3. C
4. AB

1-34. What is the bandwidth of an amplifier?
1. The actual frequencies the amplifier is effective in amplifying
2. The difference between the high and low frequencies seen at the input of the amplifier
3. The width, in inches, between the half-power points on a frequency-response curve
4. The difference between the lowest and highest frequency shown on a frequency-response curve

THIS SPACE LEFT BLANK INTENTIONALLY.

OUTPUT (VOLTS)

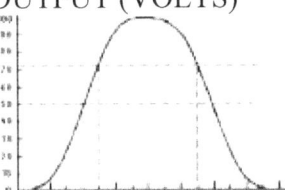

500 1.51 2-51 3.5k 4_5k 5_5k 6.5k 7.5k FREQUENCY 1k 2k 3k +k 5k Ok 7k Ok (HERTZ)

Figure 1A.—Frequency-response curve.

IN ANSWERING QUESTIONS 1-35 AND 1-36 REFER TO FIGURE 1A.

1-39. What happens to capacitive reactance as frequency decreases?
1. It increases
2. It decreases
3. It remains the same
4. It cannot be determined

1-40. What happens to inductive reactance as frequency increases?
1. It increases
2. It decreases
3. It remains the same
4. It cannot be determined

1-35. What are the (a) upper and (b) lower frequency limits shown?
1. (a) 8 kHz
2. (a) 7.5 kHz
3. (a) 6.0 kHz
4. (a) 5.5 kHz
(b) 0 Hz (b) 0 Hz (b) 2 kHz (b) 2.5 kHz

1-41. What is the maj or factor that limits the high frequency response of an amplifier?
1. Inductance
2. Resistance
3. Capacitance

4. Transformer reactance

1-36. What is the bandwidth shown?
1. 1 inch
2. 8 kHz
3. 3 kHz
4. 2 kHz to 8 kHz

1-37. Which of the following limit(s) the frequency response of a transistor amplifier?
1. The inductance
2. The transistor
3. The capacitance
4. All of the above

1-38. What type of feedback is caused by interelectrode capacitance?
1. Bypassed
2. Negative
3. Positive
4. Regenerative

1-42. What components can be used to increase the high-frequency response of an amplifier?
1. Diodes
2. Inductors
3. Resistors
4. Capacitors

1-43. What determines whether a peaking component is considered "series" or "shunt"?
1. The relationship of the component to the power supply
2. The relationship of the component to the input signal path
3. The relationship of the component to the amplifying device
4. The relationship of the component to the output signal path

1-44. What is the arrangement of both "series" and "shunt" peaking components called?
1. Coordinated
2. Combination
3. Combined
4. Complex

1-45. Which of the following components in a transistor amplifier circuit tends to limit the low-frequency response of the amplifier?
1. The transistor
2. The load resistor
3. The coupling capacitor
4. The input-signal-developing resistor

*Vcc

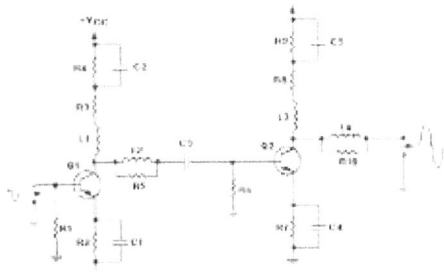

Figure IB.—Video Amplifier.

IN ANSWERING QUESTIONS 1-46 THROUGH 1-52, REFER TO FIGURE IB.

1-46. What is the purpose of LI in relation to Ql?
1. Decoupling
2. Shunt peaking
3. Series peaking
4. Input-signal developing

1-47. What is the purpose of C4 in relation to Q2?
1. Decoupling (Bypass)
2. Shunt peaking
3. Series peaking
4. Input-signal developing

1-48. What is the purpose of R4 in relation to Ql?
1. Coupling resistor
2. Input-signal developing
3. Low-frequency compensation
4. High-frequency compensation

1-49. What is the purpose of L4 in relation to Q2?
1. Coupling
2. Decoupling
3. Shunt peaking
4. Series peaking

1-50. What is the purpose of Rl0 in relation to Q2?
1. Swamping
2. Input-signal developing
3. Low-frequency compensation
4. High-frequency compensation

1-51. Which of the following components is/are used for high-frequency compensation for Q2?
1. C4
2. L3
3. R9
4. All of the above

1-52. Which of the following components is/are used for low-frequency compensation for Ql?
1. CI
2. C2
3. R3
4. All of the above

1-53. What is the effect of the gain of an amplifier if the input-signal developing impedance is decreased?
1. It decreases
2. It increases
3. It remains the same
4. It cannot be determined

1-54. What is the effect on the gain of an amplifier if the output-signal-developing impedance is increased?
1. It decreases
2. It increases
3. It remains the same
4. It cannot be determined

1-55. What is/are the purpose(s) of a frequency-determining network in an rf amplifier?
1. To create a large bandpass
2. To compensate for low-frequency losses
3. To provide maximum impedance at a given frequency
4. All of the above

1-56. Of the following networks, which could be used as a frequency-determining network for an rf amplifier?
1. A parallel-resistor network
2. A series-resistor network
3. A parallel RC network
4. A parallel LC network

1-57. Which of the following methods may be used to tune an LRC frequency-determining network to a different frequency?
1. Vary the capacitance
2. Vary the inductance
3. Both 1 and 2 above
4. Vary the resistance

1-58. What is the most common form of coupling for an rf amplifier?
1. RC
2. Resistor
3. Impedance
4. Transformer

1-59. Which of the following advantages are provided by transformer coupling?
1. Simpler power supplies can be used
2. The circuit is not affected by frequency
3. Low-frequency response is improved
4. Fewer parts are used

1-60. If a current gain is desired, which of the following elements/networks should be used as an output-coupling device?
1. An RC network
2. A resistive network
3. A step-up transformer
4. A step-down transformer

1-61. Which of the following techniques would cause a too-narrow bandpass in an rf amplifier?
1. An overcoupled transformer
2. A loosely coupled transformer
3. The use of a swamping resistor
4. The use of a frequency-determining network

1-62. Which of the following techniques would cause low gain at the center frequency of an rf amplifier?
1. An overcoupled transformer
2. A loosely coupled transformer
3. The use of a swamping resistor
4. The use of a frequency-determining network

1-63. What type of transformer coupling should be used in an rf amplifier?
1. Ideal
2. Loose
3. Optimum
4. Overcoupling

1-64. Which of the following methods provides the widest band-pass in an rf amplifier?
1. A swamping resistor
2. A loosely coupled amplifier
3. A large input-signal-developing resistor
4. A small output-signal-developing resistor

1-65.

1-66.
Which of the following methods will compensate for the problem that cause low gain in an rf amplifier?
1. Using rf transformers
2. Taking advantage of the interelectrode capacitance
3. Both 1 and 2 above
4. Using audio transformers

Which of the following types of feedback is usually caused by the base-to-collector interelectrode capacitance?
1. Regenerative
2. Decoupled
3. Positive
4. Negative

In an rf amplifier an unwanted signal is coupled through the base-to-collector interelectrode capacitance. This problem can be solved by providing feedback out of phase with the unwanted signal. What is this technique called?
1. Neutralization
2. Compensating
3. Decoupling
4. Swamping

THIS SPACE LEFT BLANK INTENTIONALLY.

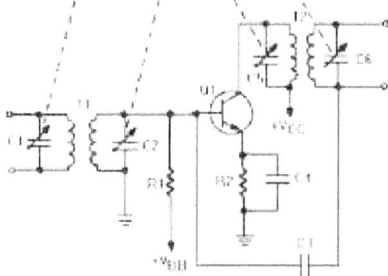

1-67.
Figure 1C—RF amplifier.

IN ANSWERING QUESTIONS 1-68 THROUGH 1-75, REFER TO FIGURE 1C.

1-68. Which of the following components is/are part of the input-signal-developing impedance for Ql?
1. CI
2. Tl
3. C3
4. All of the above

1-69. What is the purpose of Rl?
1. To provide swamping for the secondary of Tl
2. To act as an output-signal-developing resistor
3. To provide proper bias to the base of Ql
4. To develop the signal coupled by C3

1-70. What is the purpose of R2?
1. To provide swamping for C4
2. To develop the input signal for Q1
3. To provide bias to the emitter of Q1
4. To act as the output-signal-developing resistor

1-71. If C4 were removed from the circuit, what would happen to the output?
1. It would increase
2. It would decrease
3. It would remain the same
4. It cannot be determined

1-72. Which of the following components is/are part of the load for Ql?
1. C6
2. T2
3. Both 1 and 2 above
4. C3

1-73. How many tuned parallel LC circuits are shown in the schematic?
1. One
2. Two
3. Three
4. Four

1-74. What do the dotted lines connecting C1, C2, C5, and C6 indicate?
1. The components are in a different physical location
2. That the components are "phantom" components
3. The components are variable capacitors
4. The components are ganged together

1-75. What is the purpose of C3?

1. To couple the input signal of Q1 to the secondary of T2
2. To tune the parallel LC circuit of C3, C6, and T2
3. To provide neutralization for Q1
4. To bypass Rl

ASSIGNMENT 2

Textbook assignment: Chapter 3, "Special Amplifiers," pages 3-1 through 3-70.

2-1. What is the maximum number of possible inputs in a differential amplifier?
1. One
2. Two
3. Three
4. Four

2-3. If the input signals are in phase with each other, what will the peak-to-peak amplitude of the output signal be?
1. 0 volts
2. 12 volts
3. 24 volts
4. 48 volts

2-2. What is the maximum number of possible outputs in an differential amplifier?
1. One
2. Two
3. Three
4. Four

+Vcc
>R3
CI
C3
INPUT 1
C OUTPUT
Q1
C2
INPUT 2
R2
*EE

Figure 2A.—Difference amplifier.

IN ANSWERING QUESTIONS 2-3 THROUGH 2-10, REFER TO FIGURE 2A AND THE FOLLOWING INFORMATION: THE AMPLIFIER HAS A GAIN OF 6 AND THE INPUT SIGNALS ARE SINE WAVES WHICH ARE EQUAL IN AMPLITUDE AND VARY BETWEEN +2 VOLTS AND -2 VOLTS.

2-4. Which of the following statements describes the output signal if the input signals are in phase with each other?
1. A sine wave in phase with the input signals
2. A sine wave 180 degrees out of phase with the input signals
3. A sine wave 90 degrees out of phase with the input signals
4. Not a sine wave

2-5. If the input signals are 180 degrees out of phase with each other, what will

the peak-to-peak amplitude of the output signal be?
1. 0 volts
2. 12 volts
3. 24 volts
4. 48 volts

2-6. Which of the following statements describes the output signal if the input signals are 180 degrees out of phase with each other?
1. A sine wave 90 degrees out of phase with each input signal
2. A sine wave in phase with input signal number one
3. A sine wave in phase with input signal number two
4. Not a sine wave

2-7. If input signal number one is 90 degrees out of phase with input signal number two, what will the peak-to-peak amplitude of the output signal be?
1. 0 volts
2. 12 volts
3. 24 volts
4. 48 volts

Vcc
OUTPUT OUTPUT 1 2
INPUT
2
J-

2-8. Which of the following statements describes the output signal if input signal number one is 90 degrees out of phase with input signal number two?
1. A sine wave 90 degrees out of phase with input number two
2. A sine wave 180 degrees out of phase with input number one
3. A sine wave 180 degrees out of phase with input number two
4. Not a sine wave

2-9. If input number two is the only input signal applied to the amplifier, what will the peak-to-peak amplitude of the output signal be?
1. 0 volts
2. 12 volts
3. 24 volts
4. 48 volts

2-10. Which of the following statements describes the output signal if input number two is the only input signal applied to the amplifier?
1. A sine wave in phase with the input signal
2. A sine wave 90 degrees out of phase with the input signal
3. A sine wave 180 degrees out of phase with the input signal
4. Not a sine wave

-v E e

Figure 2B.—Differential amplifier.

IN ANSWERING QUESTIONS 2-11 THROUGH 2-24, REFER TO FIGURE 2B

AND THE FOLLOWING INFORMATION: ALL INPUT SIGNALS ARE SINE WAVES WITH A PEAK-TO-PEAK AMPLITUDE OF 5 MILLIVOLTS. THE GAIN OF THE AMPLIFIER IS 100.

2-11. If input number one and output number one are the only terminals used, what will the peak-to-peak amplitude of the output signal be?
1. 1 volt
2. 2 volts
3. 500 millivolts
4. 0 volt

2-12. Which of the following statements describes the output signal if input number one, and output number one are the only terminals used?
1. A sine wave in phase with the input signal
2. A sine wave 90 degrees out of phase with the input signal
3. A sine wave 180 degrees out of phase with the input signal
4. Not a sine wave

2-13. If input number one is the only input and both output terminals are used, what will the peak-to-peak amplitude of each output signal be? (Assume base of Q2 grounded)
1. 1 volt
2. 2 volts
3. 500 millivolts
4. 0 volts

2-14. If input one is the only input used and an output signal is taken between output number one and output number two, what will the peak-to-peak amplitude of the output signal be? (Assume base of Q2 grounded)
1. 1 volt
2. 2 volts
3. 500 millivolts
4. 0 volts

2-15. Which of the following statements describes the output signal if input one is the only input used and an output signal is taken between output number one and output number two?
1. A sine wave twice the amplitude of output number two
2. A sine wave 90 degrees out of phase with the input signal
3. A sine wave one-half the amplitude of output number one
4. Not a sine wave

2-16. If the input signals are in phase with each other what will the amplitude of each output signal be?
1. 1 volt
2. 2 volts
3. 500 millivolts
4. 0 volts

2-17. If the input signals are in phase with each other and an output signal is taken between the two output terminals, what will the amplitude of the output signal be?
1. 1 volt
2. 2 volts

3. 500 millivolts
4. 0 volts

2-18. Which of the following statements describes output signal number one if the input signals are in phase with each other?
1. A sine wave in phase with input signal number one
2. A sine wave in phase with input signal number two
3. A sine wave 90 degrees out of phase with input signal number one
4. Not a sine wave

2-19. If the input signals are 180 degrees out of phase with each other, what will the peak-to-peak output of each output signal be?
1. 1 volt
2. 2 volts
3. 500 millivolts
4. 0 volts

2-20. If the input signals are 180 degrees out of phase with each other and the output signal is taken between the two output terminals, what will the peak-to-peak amplitude of the output signal be?
1. 1 volt
2. 2 volts
3. 500 millivolts
4. 0 volts

2-21. Which of the following statements describes output signal number one if the input signals are 180 degrees out of phase with each other?
1. A sine wave in phase with input signal number one
2. A sine wave in phase with input signal number two
3. A sine wave in phase with output signal number two
4. Not a sine wave

2-22. Which of the following statements describes the output signal if the input signals are 180 degrees out of phase with each other and the output signal is taken between the output terminals?
1. A sine wave
2. Not a sine wave
3. A sine wave 90 degrees out of phase with input signal number one
4. A sine wave in phase with input signals number one and two

2-25. Which of the following is NOT a requirement for an operational amplifier?
1. Very high gain
2. Very high input impedance
3. Very high output impedance
4. Very low output impedance

2-26. Which of the following types of components are used in most operational amplifiers?
1. Transistor circuits
2. Electron tube circuits

3. Both 1 and 2 above
4. Integrated circuits

THIS SPACE LEFT BLANK INTENTIONALLY.

2-23. If the input amplitudes are increased to 15 millivolts and are 180 degrees out of phase, what will be the peak-to-peak amplitude of the combined output?
1. 1 volt
2. 2 volts
3. 3 volts
4. 1.5 volts

2-24. What will be the peak-to-peak amplitude of the combined output if the inputs are 6 millivolts peak-to-peak, 180 degrees out of phase, and the gain is 20?
1. 2 volts
2. 2.4 volts
3. 0.12 volts
4. 0.24 volts

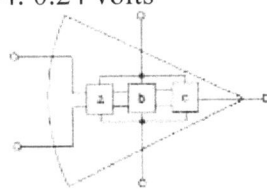

Figure 2C.—Operational Amplifier.

IN ANSWERING QUESTIONS 2-27 THROUGH 2-29, REFER TO FIGURE 2C.

2-27. What is part A of the figure?
1. An input amplifier
2. A power amplifier
3. A voltage amplifier
4. A differential amplifier

2-28. What is part B of the figure?
1. A differential amplifier
2. A voltage amplifier
3. A power amplifier
4. A video amplifier

2-29. What is part C of the figure?
1. An output amplifier
2. A voltage amplifier
3. A differential amplifier
4. A high-impedance amplifier

2-30. If degenerative feedback is used in an operational-amplifier circuit, which of the following terms describes the circuit configuration?
1. Open loop
2. Closed loop
3. Full circle
4. Neutralized

2-31. Which of the following signals determines the stability of the output signal from an operational-amplifier circuit in which degenerative feedback is used?

1. The input signal only
2. The feedback signal only
3. Both 1 and 2 above
4. The detected signal

THIS SPACE LEFT BLANK INTENTIONALLY.

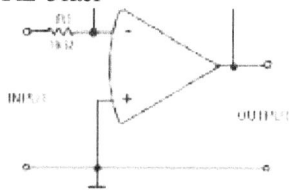

Figure 2D.—Inverting Configuration.

IN ANSWERING QUESTIONS 2-32 THROUGH 2-36, REFER TO FIGURE 2D.

2-32. In the inverting configuration of an operational-amplifier circuit where are the (a) input signal and (b) feedback signal applied?

1. (a) Inverting input (b) Inverting input
2. (a) Inverting input
(b) Noninverting input
3. (a) Noninverting input (b) Inverting input
4. (a) Noninverting input (b) Noninverting input

2-33. In the inverting configuration of an operational-amplifier circuit with feedback applied and a 1-volt, peak-to-peak, sine wave as an input signal, what is the amplitude of the signal at the inverting input of the operational amplifier?

2-34. In the inverting configuration of an operational-amplifier circuit, when the noninverting input of the operational amplifier is grounded, what is the term that describes the potential at the inverting input of the operational amplifier?

1. Feedback-signal voltage
2. Input-signal voltage
3. Signal ground
4. Virtual ground

2-35. If the amplitude of the input signal to the circuit is +2 millivolts, what will the amplitude of the output signal be?

1. -10 mV
2. -2 mV
3. +10 mv
4. +2mV

2-36. If the unity gain point of the operational amplifier is 1 mega-hertz, what is the bandwidth of the circuit?

1. 100 kHz
2. 200 kHz
3. 300 kHz
4. 400 kHz

THIS SPACE LEFT BLANK INTENTIONALLY.

1. 1 volt

2. 2 volts
3. 10 volts
4. 0 volts

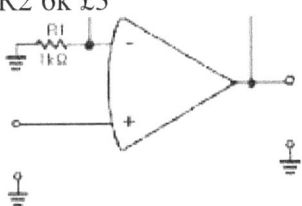

Figure 2E.—Noninverting Configuration.

IN ANSWERING QUESTIONS 2-37 AND 2-38, REFER TO FIGURE 2E.

2-37. If the amplitude of the input signal is +10 millivolts, what is the amplitude of the output signal?
1. +70 mV
2. +60 mV
3. -70 mV
4. -60 mV

2-38. The open-loop gain of the operational amplifier is 100,000 and the open-loop bandwidth is 10 hertz. If we make it a closed-loop with a gain of 10, what is the bandwidth of the circuit?
1. 100 kHz
2. 350 kHz
3. 500 kHz
4. 583 kHz

2-39. Which of the following is a difference between a summing amplifier and an adder circuit?
1. The amount of gain
2. The number of inputs
3. The type of operational amplifier
4. The placement of resistors in the circuit

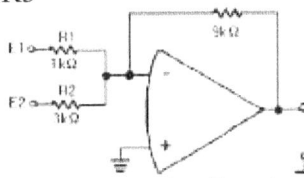

Figure 2F.—Scaling Amplifier.

IN ANSWERING QUESTIONS 2-40 AND 2-41, REFER TO FIGURE 2F.

2-40. THIS QUESTION HAS BEEN DELETED.

2-41. If the amplitude of the signal at E1 is +3 volts and the amplitude of the signal at E2 is +4 volts, what is the amplitude of the output signal?
1. +39 volts
2. +45 volts
3. -39 volts
4. -45 volts

2-42. If the amplitude of the signal at E1 is +5 volts and the amplitude of the signal at E2 is +2 volts, what is the amplitude of the signal at the inverting (-) input of the operational amplifier?

1. 0 volts
2. +7 volts
3. +21 volts
4. +54 volts

2-43. Which of the following is a difference between a difference amplifier and a subtracter?

1. The amount of gain
2. The number of inputs
3. The type of operational amplifier
4. The placement of resistors in the circuit

2-44. How many inputs can a (a) difference amplifier and (b) summing amplifier have?

1. (a) Two only (b) Two only
2. (a) Two only
(b) More than two
3. (a) More than two (b) Two only
4. (a) More than two (b) More than two

2-48. The gain of the operational amplifier shown in figure 2G can be determined by using the ratio of resistance. Which of the following ratios is correct for determining this gain.

1. $R3 = \frac{R2}{R1}$
2. $R3 = \frac{R4}{R£} \cdot \frac{R2}{R2} \cdot \frac{}{R1}$ KA~ R3
3. $R3 = \frac{R4}{R1} \cdot \frac{}{R2}$
4. (25k £2)

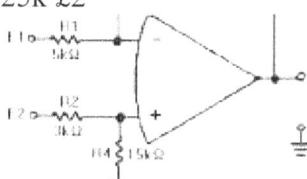

Figure 2G.—Difference Amplifier.

IN ANSWERING QUESTIONS 2-45 THROUGH 2-48, REFER TO FIGURE 2G.

2-45. THIS QUESTION HAS BEEN DELETED.

2-46. If the amplitude of the signal at E1 is +3 volts and the amplitude of the signal at E2 is +9 volts, what will the amplitude of the output signal be?

1. +6 volts
2. +12 volts
3. +30 volts
4. +60 volts

2-47. If the amplitude of the signal at E1 is +6 volts and the amplitude of the signal at E2 is +2 volts, what will the amplitude of the output signal be?

1. +8 volts
2. +20 volts
3. -20 volts
4. -40 volts

2-49. A magnetic amplifier can be classified as which of the following types of amplifier?
1. RF amplifier
2. Audio amplifier
3. Video amplifier
4. Voltage amplifier

2-50. Which of the following statements describes the basic operating principle of a magnetic amplifier?
1. Any power amplifier will create a magnetic field which can be used to increase the gain of the power amplifier
2. The inductance of an air-core inductor will change as the power used by the load changes
3. A changing inductance can be used to control the current in a load
4. Magnetism can be increased (amplified) by changing the voltage amplitude

2-51. What happens to the true power in a series LR circuit if the inductance is decreased?
1. It increases
2. It decreases
3. It remains the same
4. It increases initially and then decreases rapidly

2-53.
If the permeability of the core of a coil decreases, what happens to the (a) inductance and (b) true power in the circuit?
1. (a) Increases
2. (a) Increases
3. (a) Decreases
4. (a) Decreases
(b) increases (b) decreases (b) increases (b) decreases

If the current in an iron-core coil is increased to a large value (from the operating point) what happens to the permeability of the core?
1. It increases
2. It decreases
3. It remains the same
4. It increases initially and then decreases rapidly

2-54. If two coils are wound on a single iron core, a change in current in one coil (a) will or will not cause a change in inductance and (b) will or will not cause a change in current in the other coil.
1. (a) Will
2. (a) Will
3. (a) Will not
4. (a) Will not
(b) will (b) will not (b) will (b) will not

Figure 2H.—Saturable-core reactor.
IN ANSWERING QUESTION 2-55, REFER TO FIGURE 2H.

2-55. What portion of the schematic diagram indicates a saturable core?
1. a

2. b
3. c
4. d

2-56. A magnetic amplifier should be operated on what portion of the magnetization curve?
1. The positive peak
2. The negative peak
3. The mid-point
4. The knee

2-57. A toroidal core is used in a saturable-core reactor to counteract which of the following effects?
1. Hysteresis
2. Copper loss
3. Both 1 and 2 above
4. The effect of load flux on control flux

2-58. Why is a rectifier used in a magnetic amplifier?
1. To decrease current
2. To eliminate hysteresis loss
3. To increase the power-handling capability
4. To convert the magnetic amplifier from an a.c. device to a d.c. device.

2-59. What can be used to set a magnetic amplifier to the proper operating point and leave the control winding free to accept input signals?
1. A filter
2. A bias winding
3. A d.c. power source
4. A feedback network

2-60. A magnetic amplifier would not be used in which of the following devices?
1. A servo system
2. A d.c. power supply
3. Temperature indicators
4. A wideband audio power amplifier system

www.ingramcontent.com/pod-product-compliance
Lightning Source LLC
Chambersburg PA
CBHW081151180526
45170CB00006B/2033